Situational Awareness for Emergency Response

SITUATIONAL AWARENESS

FOR EMERGENCY RESPONSE

RICHARD B. GASAWAY, PHD

Chief Scientist
Public Safety Laboratory

FIRE ENGINEERING
• BOOKS •

Disclaimer: The recommendations, advice, descriptions, and the methods in this book are presented solely for educational purposes. The author and publisher assume no liability whatsoever for any loss or damage that results from the use of any of the material in this book. Use of the material in this book is solely at the risk of the user.

Fire Engineering Books
110 S. Hartford Ave., Suite 200
Tulsa, Oklahoma 74120 USA

800.752.9764
+1.918.831.9421
info@fireengineeringbooks.com
www.FireEngineeringBooks.com

Marketing Manager: Amanda Alvarez
National Account Manager: Cindy J. Huse

Director: Mary McGee
Managing Editor: Marla Patterson
Production Manager: Sheila Brock
Production Editor: Tony Quinn
Book Designer: Susan E. Ormston
Cover Designer: Karla Pfeifer

Library of Congress Cataloging-in-Publication Data

Gasaway, Richard B.
Situational awareness : improving decision making in public safety domains / Richard B. Gasaway, PhD, Chief Scientist, Public Safety Laboratory.
pages cm
Includes index.
ISBN 978-1-59370-307-3
1. Public safety. 2. Decision making. I. Title.
HV675.G37 2013
363.1001'9--dc23
2013000234

Printed in the United States of America

4 5 28 27 26 25 24

Incompetent behavior . . .

Without consequence (no casualty) . . .

Leads to overconfidence . . .

And contributes to complacency and arrogance . . .

Precursors to catastrophe.

Contents

Preface

Thank you for purchasing this book and for the opportunity to share with you my research and experiences on first-responder decision making and situational awareness. The inspiration for this book is simple: I want to help you see the bad things coming in time to change the outcome. I believe if we are to improve first responder safety, we must advance the understanding of how decisions are made during high-stress, high-consequence emergency operations.

Since joining the fire service over 30 years ago I have served as a firefighter, EMT, paramedic, company officer, training officer, assistant chief, and fire chief in six emergency service organizations. During that time I have responded to thousands of emergency incidents including multiple entrapments, shootings, hostage situations, building collapses, a major interstate bridge collapse, plane crashes, tornados, floods, wildland fires, explosions, hazmat spills, confined space and high-angle rescues, single-family residential-dwelling fires, and million-plus square-foot warehouse-complex fires. The personnel under my command at these incidents ranged from 6 to more than 400.

I consider myself very fortunate. I have never had a firefighter under my command receive anything more than a minor injury. In the latter part of my career I realized that much of my accomplished safety record was based, in part, on having accumulated a vast repertoire of lived experiences that helped me make good decisions. However, as I have immersed myself into the world of cognitive neuroscience research and learned how decisions are made under stress, I now attribute my early successes to being more lucky than good. *Truly, it was only by the grace of God did I not get a firefighter seriously hurt or killed under my command*. When you think about it, that's a pretty scary admission of my own incompetence.

When I became a fire chief I was 28 years old and I had 10 solid years of experience as a firefighter, company officer, and instructor. I had accelerated my experience by taking dozens of classes and participated on every live structural burn exercise I could. But I had little understanding about the duties and responsibilities of being the commander—the person responsible for everyone else's safety. Fortunately, the first department where I served as chief was fairly busy with structure fires, entrapments, and other multiple-company incidents, and I got my baptism by fire, almost literally. As I reflect on that experience it scares me to think about how woefully underprepared I was to make decisions where firefighters could have died based on what I was telling them to do. I think the only thing that saved me was the firefighters and officers I worked with were experienced enough to not let me put them in harm's way.

In addition to my 20 years of command experience, I also had the incredible opportunity to extensively study situational awareness on the journey to earn a doctor of philosophy degree. That was a very interesting and rewarding journey, and it directly ties into how I was inspired to write this book. As with nearly every doctoral program, a requisite for my graduation was to conduct *original research*. Original research requires a student to develop a proposal and investigate a subject in a way that results in *new* knowledge. In other words, simply reading content and writing a thesis about the lessons learned from other people's research would not suffice. Doctoral researchers must design and conduct a scientific study under the strict oversight of a dissertation committee and within the confines of federal regulations that govern research that involves human participants.

One piece of advice that my mentor gave me early in the program was to pick a topic about which I was passionate because I would spend a lot of time designing and conducting my research, and if I wasn't passionate about the topic there was a risk that I would burn out and quit. Given the time commitment it took for me to develop my proposal, conduct my research, and write my dissertation, this proved to be incredibly valuable advice.

Using that advice, it didn't take me long to pick my dissertation topic because it is something I had been passionate about for a very long time. For many years I had been a student, instructor, and participant in making emergency incident decisions. I had always been intrigued about how commanders made split-second life and death decisions under stress. Over the years I read case studies and reports where firefighters had been seriously hurt or killed at emergency scenes, and I found myself asking repeatedly, "Where was the person in charge?" and, "How could the commander not see this tragic event coming?"

Now it's always easy to second-guess the decision making of a commander on Monday morning (i.e., after all the facts have been gathered and the outcome has already played out). That's not fair, and it wasn't something that I was interested in doing. But I was interested in understanding what was happening at the command level during the incident. What challenges was the commander facing that may have adversely impacted his situational awareness and subsequently led to flawed decisions that resulted in a near miss or a catastrophic outcome?

So this became my quest and the focus of my research—to understand as much as I could about how decisions are made under stress; to identify the barriers that challenge a commander's situational awareness; to develop some best practices and tools that commanders could use to improve their situational awareness and decision making; and finally, to share this information with experienced and developing emergency scene commanders.

This book is designed in three parts. Think of it much like the main structural components of a house. It starts with building a strong foundation. Then walls are added and finally the roof. For you to get the maximum benefit from the recommendations and best practices contained within these pages, it's beneficial for you to have a fundamental understanding of decision making and situational awareness. This is covered in part I.

Part II explains the original research that I conducted on the topic of command decision making and, more specifically, the barriers that challenge a commander's situational awareness. As you read emergency services trade journals, it's not difficult these days to find articles where authors share their opinions and experiences about decision making and situational awareness. There's nothing wrong with that; in fact, there is a great deal you can learn from the experiences of seasoned commanders. But, in many cases, there's little or no scientific evidence supporting their claims, findings, or advice. In fact, when I started to conduct my research on situational awareness, it became painfully evident to me how little *real* research had been conducted on this topic in the emergency services field.

Part II also explores the barriers that challenge a commander's decision making and situational awareness. In other words, what is it that keeps commanders from being able to see the bad things that are about to happen? Because I believe in my heart that if commanders were able to see a bad situation developing, they would make different decisions and avoid the catastrophic outcome. Some may choose to logically deduct that it's rapidly changing, unpredictable conditions that cause catastrophic outcomes. And, in some circumstances, I agree. But the evidence is overwhelming that in many more circumstances, there were lots of advance warning indicators that the situation was going bad. Unfortunately, those warning indicators were not detected, not comprehended, or were ignored. We look at all three of these culprits in the chapters to come.

Part III contains lessons learned and best practices for commanders. The lessons learned are drawn from my experiences commanding emergency scenes, from my original research, and from numerous other research studies that have evaluated decision makers in other high-risk, high-stress, high-consequence environments like commercial airline and jet fighter pilots, tank and battleground commanders, surgical and emergency room teams, police officers, and more. Many of the best practices are drawn from my experiences and the experiences shared with me by countless commanders I have interviewed over the years.

As you read this book, I ask that you set aside your biases, beliefs, assumptions, and perceptions about emergency incident decision making. You don't have to agree with everything that I say; however, if you read this book with an open mind, you're much more likely to glean the valuable lessons that can help improve the safety of first responders in your organization.

I also want to make a note about the role of repetition in learning. Throughout this book you will see key points and lessons repeated. This is not an editorial oversight. It's done with purposeful intent. I cannot emphasize enough the role of repetition in the process of learning, and I speak to this throughout the book as I tie various lessons together. Don't let the fact that you've read something earlier disabuse you of its importance. If I repeat it, it's not only important, it's critical to ensuring your understanding of the lessons I am conveying.

Acknowledgments

A project of this magnitude requires the commitment, dedication, and support of many people. While it is my name on the cover, I acknowledge that there are many others who should share equally in the credit for this work.

Thank you to Fire Chief John Buckman for being a friend and mentor throughout much of my career, and thank you for being so generous with your amazing collection of photography. Thank you Chief Todd Johnson (and the members of the Woodbury Fire Department), Chief Chip Comstock (and the members of the Western Reserve Fire Protection District), and Sara Rolland (SR Photography) for providing many of the photographs used in this book. The lessons were enriched and the book made more enjoyable to read through your generosity.

Thank you to the many unnamed first responders who participated in my research over the past ten years and the many more who have contributed to my understanding of situational awareness through formal and informal interviews of your experiences. I know I ask difficult questions and persist for thorough answers. Your patience and sharing have provided me with a wealth of knowledge I cannot measure but I promise to pay forward.

Thank you to the PennWell and Fire Engineering team for the opportunity to share my knowledge through this book, your website, your magazine, and the Fire Department Instructors Conference—Chief Bobby Halton, Diane Feldman, Marla Patterson, Cindy Huse, Amanda Alvarez, Tony Quinn, and Mark Haugh—you were all so helpful to me and so encouraging.

Finally, and most importantly, to my family—Carol, Cameron, Hanna, Jordan, and Micah—all that I do and all that I have become has been inspired by you. While my workload is heavy and my travels extensive, I value every moment we share together. Know that my love for improving first responder safety pales in comparison to my love for you.

"Change your thoughts and you'll change your world."

—Norman Vincent Peale

PART 1

DECISION MAKING AND SITUATIONAL AWARENESS

CHAPTER 1

INTRODUCTION

First responders have a challenging and dangerous job. It doesn't matter if it's a structure fire, a hostage situation, a multiple victim shooting incident, or a hazardous materials spill; responders are charged with making decisions that can affect their own safety as well as the safety of civilian victims (figure 1–1). It's stressful work, full of tremendous complexity. The success of the incident depends upon, in part, on responders being able to make good decisions under stress.

Fig. 1–1. Providers of emergency services are often in situations where they must make high-stress, high-consequence decisions in dynamically changing environments.

Emergency scenes can be inherently chaotic. Incidents are often fast paced and continually changing. Responders can be barraged with information, some of which can be incomplete or inaccurate, yet responders are expected to make sense of it all in a very compressed period of time and make quality decisions.

The information a responder needs to process during a structural fire response may include:

- Building construction
- Structural integrity
- Fire location
- Fire direction of travel
- Smoke color
- Smoke volume
- Smoke velocity
- Smoke density
- Smoke direction of travel
- Victim assessment
- Victim location
- Victim survivability profile
- Wind direction
- Ingress for crews
- Egress for crews
- Water supply needs
- Hoseline size and deployment
- Ventilation location and methods
- Exposure protection
- Rapid intervention routes
- Salvage needs
- Overhaul
- Rehab
- EMS
- Routes of access for apparatus
- Water runoff/environmental concerns
- Media/press releases

If your public safety position is not structural firefighting, you can easily make your own high-risk, high-consequence, emergency incident decision list.

Unfortunately, as experienced decision makers know well, outcomes are not always successful (figure 1–2). When a first responder dies at an emergency incident scene, the investigation reports often look at the organization's safety policies and procedures, the adequacy and maintenance of equipment, the training level of the responders, and whether the incident was being managed to nationally accepted standards and best practices.

Fig. 1–2. A firefighter line-of-duty death causes members and investigators to ask hard questions about why the casualty happened and how it could have been avoided.

However, it is far less common for the investigation team to take an in-depth look at the decision-making processes used and the situational-awareness barriers that may have impacted the quality of the decisions. This is very unfortunate because behind every incident scene, responder fatality, or near-miss event is a decision maker charged with making and communicating strategic and tactical plans with the first priority being the safety of responders.

The first step in understanding how to improve your emergency incident decision-making skills is to understand how quality decisions are made in high-stress, high-consequence environments. To help achieve that objective, Chapters 2 and 3 describe how research in decision making resulted in two models that are used in business, industry, medicine, military, aviation, and public safety.

Adopt the Mind-set of a Student

As you read a near-miss report, a casualty report, or this book, I encourage you to adopt the mind-set of a student. As you read about how things could have gone wrong (near-miss reports), how things did go wrong (casualty reports), and how to prevent things from going wrong (this book), keep an open mind. To help you, I encourage you to ask certain questions and avoid others.

The right question to ask when trying to learn about how things go wrong is, "Why did what they were doing, at the moment in time when things went wrong, make sense to them?" This question helps you step into the shoes of the responders (and perhaps into their minds, or at least their mind set) for the moment, attempting to understand how they saw (and interpreted) the situation in the moment. It's far too easy for us to look at the outcome of an event and see much that may not have been apparent to the responders working in the proverbial heat of the battle.

The question to avoid asking is, "How could they be so stupid?" This question, and any derivative of it, is rooted in judgment of the responders. This attitude causes us to make an immediate comparison of our knowledge, skills, and abilities to theirs and, subconsciously, we're judging them as inferior. For a very long time I did this myself. I judged hundreds of responders (unjustly) as being undertrained, incompetent, and foolish. Then something happened that changed it all for me.

I was reading a line-of-duty death report that described the catastrophic outcome of some seemingly poor decisions made by first responders. My typical reaction, having read so many of these reports over the years, was to ask, "How could they be so stupid?" What they did seemed incomprehensible, even foolish. I dismissed the responders as being poorly trained and inexperienced—far different from the perception I had of myself (highly trained and highly experienced). In my mind I had come to the conclusion that I would *never* do the dumb things they had done. Case closed.

But then I learned a little more about the responders who died. They were well trained and very experienced. The incident was similar in nature to many they had responded to in the past. This got me wondering. If something like this could happen to them, could it happen to me? That was a sobering thought. Not only could it happen to me, but I now saw for the first time that I had responded to incidents where I had simply been lucky.

Lucky is not how any responder wants to feel after gaining more than 20 years of training and experience. I'd rather think of myself as good, not lucky. But the truth was, I had been lucky. I wasn't comfortable in that mind set. So I had two choices. I could continue to live in my denial that I was vulnerable to making poor decisions, or I could set myself on a new course of learning. I chose the latter, and it changed my life. Dare I say, it may have saved my life, and it may have even saved the lives of the responders I supervised at incident scenes.

The judging mind is not a learning mind. For decades I had a judging mind, and I wasn't learning how to do anything but judge others. I might add that my judgments were often harsh and unfair, because in reality I knew so little about those I was passing judgment on and so little about the actual circumstances they were facing at the moment things went bad. That's all changed now.

I no longer judge anyone anytime for anything they do that results in a casualty or near-miss event. Instead, I now ask the right questions in my quest to be a student and understand why things made sense to the responders at the moment things went wrong. I encourage you to do the same. We owe it to the responders who die in the line of duty to learn everything we can about how and why things happened so we can prevent the same circumstances from being repeated.

Here are examples of some questions I now ask that helps frame my mind set when looking at near-miss and casualty events:

- What were the responders trying to accomplish at the time things went wrong?
- What was the overall mission of the incident, and what role were the responders playing in advancing the mission when things went wrong?
- Why did it make sense for the responders to be doing what they were doing at the moment things went wrong?
- What task or objective were the responders focused on that kept them from seeing the bad outcome on the horizon?
- Were there any distractions or interruptions that drew the responders' attention away from their task?
- Did the responders understand the overall incident goals and objectives, their roles in the mission, and how other individuals/teams were going to influence the outcome?

This list is far from complete. In fact, I keep asking questions until things begin to make sense. Sometimes the questions can't be answered as comprehensively as I'd hope because the responders who hold the answers were the victims. For me, asking the right questions (and not judging) has opened my mind to learning so much. So even when I cannot interview the victims, I am now able to infer a lot about how events were unfolding around them. That is valuable. It helps me glean (and share) lessons from a different mind-set—the mind-set of a student. (See figures 1–3 and 1–4.)

Fig. 1–3. Decision researchers improve our understanding of how high-risk, high-consequence decisions are made by observing commanders during emergency operations.

Fig. 1–4. Commanding emergency incidents is stressful, especially when decisions are made that can put the lives of responders at risk.

Honor the Fallen

I have a tremendous respect for those who are willing to lay down their lives to save the life of a stranger. In my mind there is no greater sacrifice someone can make. We owe it to our fallen to learn from tragic outcomes. I am deeply saddened by the number of times I visit a public safety organization years after they've suffered a line-of-duty death and ask what they learned from their loss. I ask what they are doing differently now. Sadly, the response is silence or a dejected "nothing's changed."

In some instances I'm told the department's administrators didn't want to dishonor the fallen by asking the tough questions about what and how things happened. They felt the hero status of the fallen would somehow be diminished if questions were asked about the decision making of the fallen. I propose a perspective that is opposite from what I just described. Honor the fallen by asking the hard questions. The greatest respect we can pay to someone who gave his or her life in the line of duty is to promise that we'll do everything we can to make sure no one else from our department, or any department, should die the same way. Honor your fallen by capturing the lessons to be learned, sharing the lessons with others, and changing how your organization operates to improve safety.

CHAPTER 2

TRADITIONAL DECISION-MAKING PROCESS

In the early days of decision-making research, most of the studies were conducted in college classrooms and laboratories that provided convenient venues and ready access to participants. However, these settings were nothing like the high-risk, high-consequence environments faced by public safety providers.

Early Decision Research

This section presents an example of how an early decision research study might have been conducted. A college professor/researcher received a grant to study decision making and created an experiment that involved stock trading. The researcher used students from his economics class because they were easily accessible and would work for little money, sometimes participating for the opportunity to earn extra credit (figure 2–1).

Fig. 2–1. Most early decision-making research involved college students who lacked expertise about the subject matter they were being asked to make decisions about.

The researchers distributed packets of information on 6 to 10 companies, including historical data, biographies of key executives, product/service information, financial statements, and stock price trends. Each student then received a hypothetical amount of money to invest—say $10,000—and were instructed to evaluate each company, and then decide in which companies they wanted to invest their money. The students were allowed to buy and sell throughout the semester, keeping track of the transactions and the processes they used to make their decisions. At the end of the term the researcher asked the students questions about which companies they invested in, how much they invested, how often they traded, how much money they earned, and how they made their decisions. Whereas the student may have thought the goal of the study was to have the most money at the end of the term, the researchers' goals were to understand *how* the students made their purchasing and trading decisions.

I can't say for sure that I was actually a victim of such experimentation, but I can vividly remember participating in an exercise similar to this in one of my college economics classes. Maybe the reason I remember this so well is the professor rewarded the winner with a pizza party. I figured *what the hell* . . . I am a financially struggling 21-year-old student. I'd have done just about anything for free food. The researcher got what he wanted, and the winning student got what he or she wanted. I know inquiring minds will want to know . . . No, I didn't win. But it was research similar to this, repeated over and over again in college classrooms and laboratories, that yielded the traditional decision-making model.

The Traditional Decision-Making Process

The traditional decision-making process is sometimes referred to as the *rational decision-making process* because the process follows a rather regimented series of steps. The process is also sometimes referred to as the *analytical decision-making process* because of its heavy reliance on the use of metrics to weigh and evaluate options. For simplicity, I call it the *traditional decision-making process*.

The steps of the traditional decision-making process include:

1. Define the problem.
2. Identify the decision-making criteria.
3. Allocate weights to each of the criteria based on importance.
4. Develop alternatives for solving the problem.
5. Evaluate each alternative based on the weighted criteria.
6. Select the best alternative that maximizes the desired outcome.
7. Evaluate the effectiveness of the action that was taken.

Step 1: Define the problem

The first step in the traditional decision-making process is to *define the problem*. This makes sense because if you don't have a problem to solve, you don't really have to make a decision. Defining the problem is an important step. If you define the problem incorrectly, you vastly increase your chances of making a poor decision, so this is a step you want to think about. It's easy to get lulled into focusing on the symptoms of the problem versus defining the root problem.

Take, for example, a person having chest pains and shortness of breath. The pain and difficulty breathing are not the problem; they are symptoms of the problem. What, exactly, is causing the problem cannot be uncovered without additional investigation and analysis. Emergency medical services (EMS) personnel are taught that if a patient displays certain combinations of symptoms (e.g., chest pain, shortness of breath, pale and clammy skin, ST-segment elevation on the EKG), the patient receives a certain treatment modality congruent with heart attack as the root cause. It's the safest default treatment option for field personnel with limited diagnostic tools. However, a heart attack cannot be diagnosed in the field. We often treat symptoms, not root causes.

Step 2: Identify decision criteria

The second step in the traditional decision-making process is to *identify the decision criteria*. In this step you set up all the criteria you need to help you make the decision. Using the stock picking scenario I discussed earlier in this chapter, the various criteria you consider might include how long each company has been in business, how profitable each company has been over the past ten years, how qualified each company's management team is, and how well each company's stock has performed over the past five years.

In the public safety arena, let's say you are a police officer on patrol and you observe a driver weaving on the roadway. As you stop the vehicle you observe that the driver's behavior seems a bit out of the ordinary. The driver's breath smells fruity; his speech is slurred and he stumbles when getting out of the vehicle. Everything I've described up to this point are symptoms, not the root problem. And although you might conclude the operator is driving while intoxicated (the root problem), it very well could be that the driver is diabetic and is having an issue with blood sugar regulation. Further investigation can help you analyze the root problem. And, regardless of the root cause, you can surmise, with certainty, that the driver's ability to make good judgment is impaired and he cannot drive in that condition. But the additional assessment helps you determine whether to arrest the driver or call an ambulance.

Step 3: Weight the criteria

The third step in the traditional decision-making process is to *assign weights to the various criteria*. In other words, how important is each of the decision criteria you set in the previous step? Using the stock purchase example, is the profitability of the company more important than the quality of the management team? Is the length of time the company's been in business more important than how well the stock has performed over the past five years? Are all the criteria equally important?

Step 4: Develop alternatives

The fourth step in the traditional decision-making process is to *develop some alternatives* (choices). If you are a stockbroker, the list of alternatives includes all of the companies traded on the stock exchange. However, as you can imagine, there are thousands of publicly traded companies, and it would be impossible to consider and apply the criteria and weighting method to every company. So the trader might limit the alternatives by applying some additional criteria.

For example, the trader might want to only choose companies traded on the New York Stock Exchange; or the trader might want to choose companies who are considered blue chip; or the trader may want to consider companies that are considered mid-level capitalized. Don't get knotted up about knowing what all these terms mean. You're not going to be asked to pick stocks. Just know you may want to apply additional filters to get your choices down to a manageable number. How many choices are manageable? Great question. We address that in chapter 4.

Step 5: Evaluate alternatives

The fifth step in the traditional decision-making process is to *evaluate each alternative*. After you have all the alternatives you want to consider, apply the weights for the decision criteria you set up in steps 2 and 3.

This results in a numerical score for each alternative. The choices are then ranked from the highest scoring (best choice) to the lowest scoring (worst choice).

Step 6: Make the decision

The sixth step in the traditional decision-making process is to *make the decision*—to pick the highest-ranked alternative. Of all the steps, this one should be the easiest. You've applied a logical, arguably scientific, way of evaluating your choices, and the best decision has risen to the top of the list. So decide and move on. Easy enough, right?

Enter emotions

The traditional decision model is a solid process . . . until your emotions start to get involved in the process. Every step in the process up to this point has been very rational, scientific, numerical, and logical. It all makes sense. Then you find out that the company you really liked (maybe because they make a product you use all the time) didn't fare so well. Maybe that company isn't even in the top 10 based on the criteria you had set. But you've heard some great things about this company, and you have a really good *feeling* about its future performance. Now what do you do?

You're torn between your logical, rational side and your intuitive, emotional side. It can be an interesting dilemma, and what you should do about it isn't always obvious. In other words, you can't say that always making the logical, rational decision is best. Nor can you say that always making the intuitive, emotional decision is best. And finally, you can't always say that blending the two will result in the best decision. Nonetheless, you make your decision and put your plan into action.

Step 7: Reflection

This leads to the seventh and final step in the traditional decision-making process—reflection. In this step you *evaluate the effectiveness of your decision*. In other words, how did you do? Did you make a good decision or a bad decision? Did things turn out as you expected, or were there some surprises? Maybe there were consequences as well.

The clock is ticking.

A critical consideration in the public safety arena is how long you should wait before you evaluate the effectiveness of your decision. The greater the risk and consequence and the more compressed the time line (i.e., the speed in which the incident is moving), the quicker you have to assess if your decision was a good one. Change the course of action too soon and you may not have given the decision an opportunity for success. Delay changing the course of action too long, and you may find yourself drifting into a catastrophe.

Experience Is the Secret Sauce

> *I've learned more in my life from the mistakes I've made than from the things I've done right.*

Reflection is the seventh and final step mentioned in the previous section. This reflection on your decisions is the kind of stuff that experience is made from. Stated simply, the collective outcomes of your decisions and the lessons you've learned from them are what help you make better decisions.

Children (and teenagers) tend to make poor decisions because they don't have the collection of experiences that you have—the lessons from all the mistakes you made while growing up. And this is where parents (and in emergency services, *mentors*) can be so darn valuable.

You may not want to admit it, but those *old* people may actually know a thing or two about the problems you're facing. Why? Not because they're old, but because they are *wise* (figure 2–2). That wisdom is the collection of their experiences, success, failures, and the evaluation of their lessons learned along the way.

Fig. 2–2. Young firefighters benefit from the shared knowledge of seasoned veterans. The passing of knowledge through mentoring helps newer firefighters become better decision makers.

Having spent 30-plus years in emergency services, my mission now, as an old firefighter, is to ensure young firefighters become old firefighters.

I Need a Vacation

Here's an example of how to make a decision using a traditional process. The scenario revolves around a person contemplating a vacation. An application of the traditional (rational) decision-making process might help make a good decision.

Step 1: Define the problem.

– I've been working too hard. I'm burned out and I need a vacation!

Step 2: Define the decision-making criteria.

– Cost, amenities, activities, solitude, weather

Step 3: Allocate weights to each criteria based on how important they are to you.

– Cost = 50%

– Amenities = 20%

– Activities = 15%

– Solitude = 10%

– Weather = 5%

Step 4: Develop alternatives for solving the problem.

– Beach

– Mountains

– Cruise

– Theme park

– Visit relatives

Step 5: Evaluate each alternative based on the weighted criteria.

(Ranked order for first decision criteria—most expensive to least expensive)

1. Theme park $4,000
2. Cruise $3,500
3. Beach $3,000
4. Mountains $2,500
5. Visit relatives $1,000

Multiply the criteria weight by the ranking weight. For example, the weighting for expense (because that is a big concern) is 50%. So multiply the option ranking by the criteria weighting.

- Theme park: (50 × 1) = 50
- Cruise: (50 × 2) = 100
- Beach: (50 × 3) = 150
- Mountains: (50 × 4) = 200
- Visit relatives: (50 × 5) = 250

Visiting relatives gets the highest score, reflecting it's the lowest-cost alternative. Conversely, the theme park option received the lowest score because it is the highest-cost alternative. Repeat Step 5 for each of the decision criteria. The collective scoring for all the criteria and alternatives is charted in figure 2–3.

	Beach	Mountains	Theme Park	Cruise	Relatives
Cost (50)	3×50–150	4×50–200	1×50–50	2×50–100	5×50–250
Amenities (20)	4×20–80	1×20–20	3×20–60	5×20–100	2×20–40
Activities (15)	3×15–45	2×15–30	5×15–75	4×15–60	1×15–15
Solitude (10)	4×10–40	5×10–50	2×10–20	3×10–30	1×10–10
Weather (5)	5×5–25	1×5–5	3×5–15	2×5–10	4×5–20
Weighted Totals	340	305	220	300	335

Winner!

Fig. 2–3. The traditional decision-making process uses analytical decision making to evaluate alternatives. It is a very logical process.

There's Just One Problem

This is an effective process for making decisions. It's rational, and it reflects careful consideration to all the essential criteria and priorities. However, you must ask:

- Does the rational decision-making process accurately reflect the realities of how decisions are made at emergency scenes?
- Is there enough time to do something like this?
- Is there enough good data?
- Is the data stable enough for us to assess it this way?

In many instances the answer to these questions is no. Perhaps if you were able to hold the incident in abeyance for 20 or 30 minutes, you might be able to effectively apply the traditional decision-making process. A hazardous materials spill that is confined, presenting little immediate danger to life, property, or the environment, might serve as one example where time is on the side of the decision maker and the traditional decision-making process could (and some may argue *should*) be used.

However, more often, the decision maker is under an extreme time pressure to do something immediately in an effort to prevent an emergency from becoming exponentially worse. Experienced public safety personnel, under the compression of time, are not likely to use such a rational decision-making process (fig. 2–4).

But, if first responders don't have time to use a rational decision-making process, does this imply, perhaps, that they are using an *irrational* decision-making process? No, not at all—although at times their behavior may lead you to believe they are acting irrational. Chapter 3 introduces you to the alternative decision-making model that is far more realistic for making high-stress, high-consequence decisions in dynamically changing environments.

Fig. 2–4. An incident commander operating in a dynamically changing environment will struggle to make effective decisions if relying entirely on analytical decision-making processes.

CHAPTER 3

RECOGNITION-PRIMED DECISION-MAKING PROCESS

Experienced public safety commanders, making decisions in high-stress, high-consequence, time-compressed situations, often don't have time to employ the rigors of rational, analytical decision making. I'm not making excuses for public safety providers or providing an excuse for fly-by-the-seat-of-your pants decision making. There is a process used in domains of considerable risk, consequence, and time compression.

Recognition-Primed Decision Making

The process of *recognition-primed decision making* (sometimes referred to as RPD or RPDM) is an effective method for making high-risk, high-consequence decisions. This model was first discovered by Gary Klein, a cognitive decision researcher, while conducting a study with firefighters in the 1980s. Klein and several of his associates were hired by the United States Army to help improve battleground decision making (certainly a high-risk, high-stress, high-consequence, and time-compressed domain). The army had been training battleground commanders to use the traditional decision-making process (figure 3–1). Why? Because it was the only process that existed. That is, until Gary Klein rocked the decision sciences world with his RPD discovery.

Fig. 3–1. Soldiers operating in battleground conditions. (Photo by Sgt. Christopher O'Quin.)

The Birth of RPD

The researchers in Klein's group knew that the best way to understand the complexities and challenges of making decisions under stress on the battlefield would be to observe the decision makers in their natural environments (sometimes referred to in the research community as *going native*). This would entail watching battleground commanders in their natural working environment while making decisions under stress.

However, it was impractical (not to mention dangerous) for the researchers to go into battle with commanders to observe their decision making first-hand. So they went on a quest to find decision makers whose command environment closely resembled the conditions found on a battleground. That led Klein and his associates to the fire service and a study of the decision making of fireground commanders. Fireground commanders were an ideal fit. They work in time-compressed, high-stakes, rapidly changing, and unpredictable hazardous environments, and the researchers could observe them during actual emergency incidents without putting themselves in harm's way (figure 3–2).

Fig. 3–2. Fireground commanders operate in a time-compressed, high-stakes, rapidly changing, and unpredictable environment.

The goal of Klein's associates was to observe and understand how fireground commanders use the traditional decision-making model. They theorized that understanding how fireground commanders make decisions would help them help the Army solve their battleground command challenges. It was through a series of observations and interviews that Klein realized fireground commanders did not use the traditional decision-making process. It shocked him to learn that the way fireground commanders made decisions didn't even fit the standard definition of a decision (a choice between two or more alternatives). Fireground commanders didn't identify decision criteria. They didn't apply weights. They didn't compare alternatives. And they didn't select the best alternative from among a collection of choices.

Rather, experienced fireground commanders gathered a small amount of information quickly, processed what it meant, came up with a solution, and visualized an action plan to see if it would work out (termed *mental modeling*). If the outcome of the mental model was acceptable, the commander put that plan into action and never gave thought to an alternative plan as a comparison to the first plan that worked (at least initially). If the mental model of the plan the commander contrived did not work (i.e., there is a bad outcome), the commander scrapped the plan and modeled out an entirely new plan. Commanders never compared alternatives (comparing the plan that did not work with the plan that did work to see which one was better). Commanders completely disregarded the plans that did not work and considered it unproductive to compare a new plan with a plan that had already been determined to be unfeasible.

When the commanders found a plan that worked, that prompted the decision and the plan was put into action. It was a significant finding in Klein's research and a departure from what they had expected to learn—which was how fireground commanders used the traditional decision-making process. The commanders in Klein's study never considered comparing two choices side by side and selecting the best one. So, in essence, commanders were not making decisions, at least not by the traditional definition (a choice between two or more alternatives).

A decision was always thought to be a choice between two or more alternatives, but Klein discovered that was not how fireground commanders were making decisions.

Experience—A Key Ingredient

As you might imagine, to use a decision-making system that depends on a good solution popping into the commander's head can, at most, be very risky and, at the very least, be scary to the commander and those who must follow those orders. For commanders to successfully use recognition-primed decision making, they must have a level of expertise to draw upon. Having a collection of experiences and training allows commanders to recognize how to deal with problems because they have seen the same thing (or something similar) in the past and have learned valuable lessons about what has worked well (and not so well) previously.

Time compression and dynamically changing environments

One facet that separates many first responder decision environments from others is how dynamically changing the environment is, coupled with the compression of time and the need to make rapid decisions—an essential element for successful outcomes at many emergency scenes. Commanders must size up situations quickly, often in less than a minute. If a commander is going to make decisions where the consequences could cost the lives of responders and/or civilians, some might argue that only a fool would rush such an important decision. It would make much more sense for the commander to take some time, perhaps 5 to 10 minutes, to make such a critical life and death decision. But experienced fireground commanders know this is extremely difficult, if not impossible, to do. The incident conditions change so rapidly that information gathered 5 minutes ago would have degraded and assumptions made on the information gathered 5 minutes ago may no longer be valid. Too much will change in a 5-minute period. So the size-up and the processing of what is going on must be done quickly for the commander to have a comprehension of what is happening. The amount of time a commander spends on the size-up will vary depending on the complexity of the incident and the size of the incident scene.

I have posed this question to hundreds of incident commanders: "After you arrive on the scene of a single-family, 1,200 square-foot, residential dwelling fire with no exposures, how long does it take for you to do your scene size-up?" The answer is nearly universal: "Seconds." When I press them for a number, the range is typically "5 to 15 seconds," with some commanders saying, "However long it takes me to do a 360-degree size-up." I want to hug the people who say that because they understand the process of completing a size-up includes seeing as many clues and cues as possible.

Suffice it to say, most commanders don't take 5 to 10 minutes. They only take a few seconds to gather the essential information needed to make an action plan after arriving at the emergency scene.

Expertise is important

To be able to conduct a size-up and quickly make sense of what is going on and how to handle the situation requires expertise (figure 3–3). A novice decision maker, lacking experience, cannot do this very well. There's too much information to gather, process, and comprehend. The person who lacks experience has to pause and think about what each piece of information means. Experts, on the other hand, are able to comprehend the meaning of clues and cues much quicker. In fact, experts can quickly assemble groupings of clues and cues, mentally package them together, and determine the meaning of the packaged information. For example, an experienced paramedic can take a quick glance at a trauma patient and determine relatively quickly how severe the trauma is simply by packaging together audible and visual clues and cues (figure 3–4).

Fig. 3–3. Expertise allows fireground commanders to conduct rapid size-ups of situations.

Fig. 3–4. Experienced medical responders can conduct rapid triage of trauma victims by assessing a relatively small number of visual and audible clues and cues.

Novices also struggle with the task of the fitting pieces of information together to form a *big picture* perspective of what's going on. Their attention can be drawn to what appears to be important things, and they can completely miss a lot of smaller clues and cues that, alone, may not mean much individually, but collectively can mean a lot.

Negative clues and cues

Another advantage that experts have over novices is their ability to process and comprehend *negative* clues and cues. Negative, in this context, does not mean *bad* clues or cues. Negative means *absent* clues or cues. Novice decision makers can size up a situation and readily identify the clues and cues they see and hear that help them make their decision. Experts do the same thing; however, experts possess the additional ability (based on their experience level) to be able to size up a situation and comprehend the meaning of clues and cues that are *absent* (but should be there). In these situations, the absence of certain clues and cues are as critically important as the presence of others. However, because novices lack expertise, they are not able to comprehend the meaning of what is missing.

Here's an example I recall from paramedic school. One time in class the instructor asked the students if sick babies cry. Even though I was the youngest sibling in my family, I had seen enough sick kids as an EMT to answer in the affirmative. Then she asked the class if kids cry louder the sicker they are. Again I answered in the affirmative. To my surprise, the instructor informed the class that really sick kids—the ones who may be ill enough to die—oftentimes don't cry. They are too scared to cry, so they remain quiet.

Little did I know this line of inquiry was a setup for a lesson that would later pay off many times for me. If that lesson had not been taught to me in medic school, I could have easily been fooled by this negative clue. The absence of crying can be a significant indicator of critical illness or injury.

The Debate

Some could argue that it's not necessary to conduct a rapid size-up and engage first responders so quickly into action. I agree there is some logic to this argument. Nothing says you have to conduct aggressive interior structural fire attack, launch an aggressive search for victims, or barge into a house where an assailant is holed up with hostages without taking time to consider and compare your alternatives. Others would argue that to make a difference in the outcome (to save lives and property), the action required is rapid and necessary.

I recall attending a program where the lessons for public safety personnel who responded to the Columbine High School massacre incident were being assessed. One of the take-away lessons was the presenter thought the responders took too long to assess the situation, and this contributed to additional casualties. I'm not taking this position. I can't. I wasn't there to know what was the appropriate thing to do. But the presenter was critical of the decision makers who were reportedly gripped by the need for more and more and more information and, because it was constantly changing information, it took a very long time to gain a good understanding of what was really happening. It's easy to analyze and dissect decisions after the fact and be critical. I discourage the practice of Monday-morning quarterbacking. It's not productive, and it's disrespectful. It's always helpful to frame an analysis of decision making by asking: *"Why did it make sense to them at that time?"*

The novice disadvantage

Experienced commanders who possess the ability to quickly capture and comprehend information under stress are the best at completing rapid size-ups and implementing quick action plans. It can be problematic when novice commanders observe their experienced counterparts conducting rapid size-ups followed by rapid actions and think they can do the same thing. This happens because a novice is not likely to understand what the expert is doing, mentally, in those first few seconds.

When novices try to conduct rapid size-ups and take quick actions, the risk associated with their decision making increases dramatically (figure 3–5). What's worse, novices may not even know what they don't know (or realize what they missed and failed to comprehend). They are confident, but essentially incompetent. The progression of first responder competency will be discussed in greater detail later in the book.

Fig. 3–5. Inexperienced firefighters are able to see and process positive clues and cues. However, their lack of expertise can inhibit their ability to capture and process negative clues and cues.

Physical Reactions to Stress

The stress of operating at an emergency scene where the lives of responders and civilians rely on quality decisions can cause a significant amount of stress. As a first responder, you already know that. But what many responders don't fully understand is the impact of stress on the process of making good decisions. Stress contributes distinct, identifiable, and measurable physical, mental, and emotional changes including increases in blood pressure, pulse and respirations, heart palpitations, profuse sweating, dry mouth, trembling, narrowing of attention, and hyper awareness to name a few. These reactions to stress are natural and, for many, difficult to control.

Stress is a game changer

Stress triggers a reaction in the hypothalamus, a small pea-sized organ sitting approximately in the middle of your brain. This organ, in turn, sends out messages to other parts of your brain and other organs in your body that you are having a stressful event. These electronic messages trigger the release of hormones and chemicals that are designed to help you deal with the stress. Adrenalin is released to prepare you for action. Glucose is released to give you energy. Endorphins—your body's naturally produced morphine—are released to mask pain in the event you are injured. Cortisol is released to help counter the effects of the adrenaline. Suffice it to say that under stress you have a lot of thought- and performance-altering drugs coursing through your veins. These chemicals change the way you perform under stress.

Avoiding the chemical dump

The only hope you have in reducing the impact of stress is to reduce the number of chemicals being released into your body in the first place. This can be accomplished using stress-calming techniques like controlled breathing. This helps calm the hypothalamus and prevent a reaction that can have undesirable consequences. Learning how to remain calm under stress and then practicing the techniques that work for you can also reduce the release of chemicals and help you maintain some control over your reactions. Finally, the more often you are exposed to stressful events, the more likely you are to develop immunity to stress. You only need to observe the demeanor of a veteran responder to that of a new recruit at an emergency scene to witness the difference in their reaction to the stressful event (figure 3–6).

Fig. 3–6. A recruit firefighter going through basic training is likely to experience more stress and anxiety than an experienced firefighter whose been under stress many times.

When I was a new firefighter I remember my instructor telling me during self-contained breathing apparatus training to take deep breaths, hold it, and then breathe out slowly. He told me this would help slow down my breathing. He was right; it worked. What he didn't know (or at least never admitted to knowing) is the breathing technique I was being taught slowed my breathing because it calmed me down. It also helped prevent my brain and body from being overrun by stress-released hormones and chemicals. As a side note, in addition to calming my stress, slower respirations improved my air management.

Hormones and chemicals

Stress hormones and chemicals trigger primal reactions dating all the way back to your cave-dwelling ancestors. Granted, what triggered stress in a cave dweller was very different from what triggers stress in you. You stress about your kids getting into a certain college; cave dwellers stressed about their kids being eaten by a saber-toothed tiger. Nonetheless, stress is stress and the reaction is the same. Stress increases your heart rate. Consequently, to ensure the proper oxygenation of blood, your respirations increase. Your pupils also dilate to let in more light. This improves your visual acuity, making it easier to see the saber-toothed tiger if the attack occurs at night. The glucose release I mentioned earlier gives you the energy to react.

The endorphins (or more accurately endogenous morphine), while fulfilling the valuable task of masking pain, also act as stimulants; thus, they make you more excitable.

You've probably seen the effect of endorphins in action before. You respond to the scene of an automobile accident, and you see the horrific damage done to a vehicle and you know, almost instantly, anyone who was in that vehicle is either dead or at the very least seriously injured. You approach the vehicle, and you see there's no one in it. Then you notice someone standing on the side of the road who tells you they were the driver of the vehicle and they deny having injuries. How could this be? The vehicle is nothing short of a twisted mass of metal. And the driver is saying to you, "*I'm OK. I'm not really hurt.*" Then a short time later the driver starts to complain about pain in their back or neck. Set aside the fact for the moment that this could be *insurance-related* pain. Alternatively, the explanation may be when the accident happened endorphins were released into the person's bloodstream. As they are designed to do, the endorphins masked the pain of the injuries. As the endorphins worked their way through the system, the victim started to feel the pain of their injuries.

I have witnessed the phenomenon of endorphins masking pain on numerous occasions at emergency scenes, and if you haven't yet, you will. As a decision maker under stress, you will get this rush of endorphins that will make you excited and may even give you a euphoric high.

Physiological changes from stress

The chemicals released under stress cause your heart, lungs, brain, and muscles to kick into high gear. That's good because you need extra blood to provide oxygenation and nutrients to handle the stress event. Your body also restricts the blood supply to your kidneys and your digestive tract. That blood is redirected to the organs and muscle groups that are working extra hard under stress. If this did not occur, you might not be nearly as physically strong or mentally sharp in a stressful situation.

Another interesting physiological reaction that can occur under stress is the inability to control the release of contents from your bowel and bladder. Yes, it's true. If your department responds to medical calls, I can say with confidence that you have probably responded to emergency calls where someone was injured or having a significant medical crisis and lost bowel or bladder control because of stress. It's all part of the naturally occurring stress reaction.

The impact of stress on decision making

The process by which decisions are made under stress is different from how decisions are made in the normal course of daily life. This should not come as a surprise. If you're driving down the road and come upon a four-way intersection, you have to make a decision whether to go left, right, or straight ahead. Typically you've had some time to think about it. The need to make the decision was anticipated, and there was plenty of time to process information to help you make a quality decision.

Conversely, if you're driving down the road and a deer bolts out in front of your car, your decision-making environment is very different. This event is high-stress, high-consequence, and time-compressed. You didn't know it was coming and had just a nanosecond to process information before making a decision that could result in a near miss or a big messy problem.

In both scenarios you are processing information and making decisions. Just one happens slowly with low stress and low consequence, while the other is at warp speed and fraught with stress and consequence. In both cases it's just a matter of seconds to know if the decision was a good one.

The speed of a decision

Is it possible to process information and make fast decisions? You bet it is. Think about a major league baseball player up to bat and facing a 90 mile-per-hour fastball. The batter has just under a half-second from the time the ball leaves the pitcher's hands until it hits the catcher's mitt. That decision to swing (or not to) takes just slightly longer than a blink of an eye. Incidentally, the clues and cues that help the batter decide whether to swing starts well ahead of the ball ever leaving the pitcher's hand.

Making rapid decisions under stress is one of the hallmark qualities of good public safety responders, whether it's on a fireground, during a shootout, or while triaging accident scene victims. Most responders do it without really understanding the mystery behind how they make such good decisions in time-compressed environments. I'm going to take the mystery out of the process of how you do this and why you are so darn good at it.

Decisions under stress

The findings in neuroscience support the notion that stress can change behavior and the way you make decisions. Your natural, rational, analytic judgment system does not function the same under stress. For a variety of reasons, some relating to hormones, some relating purely to time constraints, your rational judgment system is not terribly effective at making split-second decisions. But wait! If you're not thinking rationally, are you thinking irrationally? You might think so sometimes as you observe the seemingly irrational behavior that some responders display under stress. But, rest assured, the behavior is not irrational. In fact, it's quite predictable.

As your stress level increases and your level of personal risk goes up (perceived or real), primitive instincts kick in and you start thinking (and sometimes acting) like your cave-dwelling ancestors. I'll speak to that in a little more detail shortly.

Most people residing in civilized nations dwell in a rational decision-making environments—where in the normal course of day-to-day living one is far more likely to make rational, analytical decisions based on easily accessible facts and data. Decision makers can easily identify an array of options and logically deduce which one is best for the situation.

Under normal (i.e., low stress) conditions, you are able to gather and comprehend complex information and apply analytical techniques that help you make decisions. As your stress level increases and your body triggers the physiological reactions discussed earlier, it becomes more difficult to comprehend complex and detailed information. You may even struggle to understand things that, under normal stress levels, you would find easy to comprehend. For example, here is a word problem I borrowed from one of my son's third-grade homework assignments.

> There are three playing cards lying face up, side by side on a table in the following configuration: A five is just to the right of a two. A five is just to the left of a two. A spade is just to the left of a club, and a spade is just to the right of a spade. What are the order and suite of the cards on the table?

Figure 3–7 contains the solution to this word problem. Under normal stress levels (which I hope you are under as you read this book), and with the aid of a deck of cards to assist you with the visualization, you may be able to solve this problem relatively easily. However, if you were under stress of risk and consequence and some of the information was changing as you were trying to solve the problem, you may struggle to make sense of what is happening and how to solve the problem.

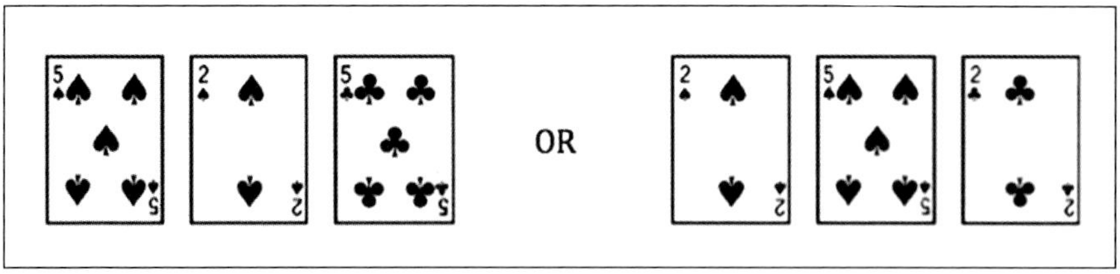

Fig. 3–7. Solution to the third-grade card riddle

Imagine this scenario: You are part way through solving the previous word problem and you're told that there was a mistake in the original data—the five is actually a seven, and the spade is actually a heart. You might find this scenario confusing and frustrating (which can create stress), and you might find yourself struggling to understand, "*Why are these cards changing? Why can't I just solve the original problem?*" And saying to yourself (out of frustration), "*I'm not going to try to solve this until I know the cards aren't going to be changed on me any more.*" If you were given this word problem to solve in a calm and controlled setting (and didn't have the information changed while you're trying to solve the problem) I am confident you'd be able to figure it out; however, if you were put under stress with added risk and consequence, experienced a change in information—and for good measure, added external environmental stressors such as bright lights and obnoxious noises—you might find yourself struggling to solve this problem. Why? All the distractions and changing conditions add stress.

Stress drives a person to use a primitive, instinct-driven form of decision making, one that draws upon mechanisms that have been genetically engineered into the human brain over millions of years of evolution. Intuition, triggered by stress, can lead to the often-referenced intuitive *gut feelings*, a primal decision process that uses subconsciously stored training and experiences supported by prewired evolutionary programming in your brain.

The physical reactions to stress have been well studied and documented by researchers and scientists. The term often used to describe the physical reaction to stress is *fight or flight syndrome* or *fight or flight response*.

Earlier I stated that when you are under stress, your rational decision-making process is impacted and you revert to using an intuitive process that can result in a gut feeling about what is going on and what you should do. Without understanding the science behind how this works, it could be a little unnerving to think an emergency scene decision maker might simply rely on gut feelings to make critical life-and-death decisions. Some might argue it defies logic and it's irresponsible to rely on intuition. It's a common argument in decision research circles, especially among those with strong convictions for rational or intuitive decision processes. There is a number of research studies that have shown decisions made using a rational decision process yielded no better outcomes than decisions based on intuition.

Intuitive decisions draw on your collective life's experiences stored in the subconscious mind; therefore, it's important to spend a little time to ensure you understand the role of intuition in decision making. This will allow you to be comfortable and confident with your intuitive decision-making abilities.

Memory

To help demystify the intuitive decision-making process, I use a few analogies and metaphors. The first analogy is *your brain is like a computer* (figure 3–8). Those versed in neuroscience, admittedly, do not favor this comparison because your brain is really nothing like a computer. Notwithstanding the criticism I may suffer from the scientific community, it really does make explaining some components of brain function easier so I'm willing to take the risk.

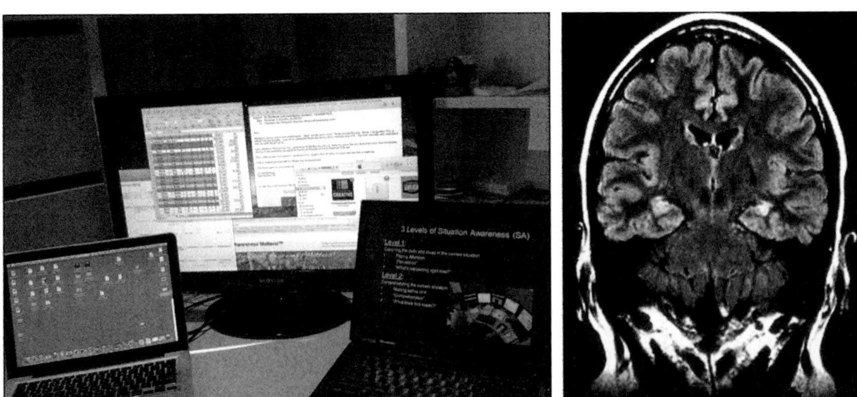

Fig. 3–8. The brain is often compared to a computer, which can be a dangerous analogy because the two operate very differently. However, the metaphor can be useful in explaining some functions.

Your computer has memory, both short-term (random access memory or RAM) and long-term memory (the hard drive). Your computer's short-term memory is only a fraction of its long-term memory capacity. Short-term computer memory allows the user to draw up programs and information from the long-term memory and put them in a space where they can be readily accessed and used. When the need for that short-term memory data is complete, it goes back into the long-term memory until it is needed again. (My apologies to the tech-savvy readers who can readily tell I am simplifying how computers work.) The user of a computer determines what programs and information are stored on the hard drive, where it is stored, and assigns a name to each file. The user can even search for the files and data using keywords. Files and programs can also be deleted from the computer's hard drive.

In some respects your brain resembles a computer. It has the ability to store and retrieve information. It stores different types of information in various places. It has a short-term memory (the neuroscience term is *working memory*) and it has a hard drive (the neuroscience term is *long-term memory*). Working memory holds the things you are thinking about or working on in the current moment in time. It's horribly vulnerable to degradation if the data is not quickly passed along to long-term memory stores.

The short-term memory can be populated with information extracted from long-term memory stores, or it may be new (novel) information being received and processed (think of this as learning). Like a computer, your short-term memory is a fraction of the size of your long-term memory, and like a computer there is a limit to the capacity of your short-term memory.

When the capacity of the short-term memory is exceeded, bad things can happen. This is catastrophically important to understand because your short-term memory is relatively small, and the consequences of short-term memory overload in public safety domains can be horrific.

If you've ever had the pleasure (a term I use loosely) of overloading the limited random access memory (RAM) on an older computer by trying to run a newer, RAM-intensive program, then you know what can happen when your brain gets overloaded. The program may run very slow, stutter (appear to skip and hop), or completely freeze up. When the information being processed exceeds the capacity of the computer's RAM, performance suffers. The same thing can happen when your brain's working memory capacity is exceeded.

Indisputably, your brain has a limited amount of short-term memory available to capture and process information. When the information coming in exceeds your brain's capacity to process it, your ability to make decisions may slow down, stutter (skip and hop around), or your brain may just freeze up, just like a computer does. Scary, huh?

First responders have shared many first-hand examples with me of this very thing happening to them under stress. They were overwhelmed with information coming into their short-term memory at an incident scene, and it became impossible for them to process and comprehend everything that was happening. I discuss this challenge in greater detail later on.

There are a few differences that separate how your brain works from how your computer works. With your computer, you get to decide the names of your files and assign the storage location for those files. If your computer starts to run slow, you can selectively close programs and shed information from the RAM cache. This can be done in whatever fashion or priority that makes sense for you.

Unfortunately, your brain doesn't work the same way. Imagine how much easier life would be if it did. Imagine if someone inquired about what you ate for lunch on December 1, 1998. Chances are, unless that date holds some special significance for you (e.g., it might have been a milestone birthday and you ate something special that day), you are not likely to remember. But it's very likely the details of what you ate for lunch that day are stored on your hard drive (or more accurately, in your long-term memory). Research has proven that much of what you see, hear, smell, taste, touch, and experience is captured and stored in memory. But you don't get to decide what is kept and what is thrown out. Nor do you get to decide where it's stored or how easily it's recalled if you need it.

Storage

For the sake of simplicity, you can think of those experiences stored in your memory as similar to photographs, cataloged into albums, representing your collective life's experiences. In reality, it's a little more intricate than I've portrayed here. Your brain actually stores various component parts of lived experiences in different places in your brain and stores the information for recall in some very complex ways.

Again, thinking about it simplistically, your brain stores smells in one area, tastes in another, visual images in another, and sounds in another. It even stores vertical lines in an image in different places than it stores the image's horizontal lines.

Stimuli can then trigger vivid and instant recalls of experiences. For example, a particular smell (say, the smell of chocolate chip cookies baking) might instantly trigger memories of your childhood visits to your grandma's house at Christmas time. This is not one memory stored in one place. Rather it is a complex recall of many memories stored in many places across your cortex. Researchers haven't been able to completely crack the code on how your brain pulls it all off so quickly and accurately, but it does.

In one study, a psychologist was able to hypnotize an adult patient and the patient was able to recall memories dating to back to very early childhood, including his first birthday party. The patient vividly recalled the guests present, the gifts received, and the flavor of the cake and ice cream (figure 3–9).

Imagine how large the hard drive capacity of your brain would need to be to store so much detail about so many experiences. I recall reading where a neuroscientist once predicted the capacity of the human brain as being greater than all the information currently residing on the Internet. It amazes me to imagine it.

Fig. 3–9. Under hypnosis, psychologists have been able to get adults to recall as far back as their first birthday party.

Memory loss

Some neuroscientists have suggested the only way to lose information stored in your brain is to damage your hard drive—to suffer brain damage, either from an injury that physically damages the anatomical structure or from diseases that alter or destroy brain function. However, researchers also acknowledge that brain cells are continually destroyed and regenerated throughout the normal course of life. Your brain also deploys some coping mechanisms that can suppress traumatic and stressful memories. We can all be thankful for that; otherwise, there might be many more neurotic people among us.

Recall

It may be a little unsettling to learn you have so little control over where and how your experiences are stored. And, as if that was not disconcerting enough, for the most part you have little conscious control over how your brain recalls much of the information you may need.

There are many factors that contribute to your ability to readily recall information, such as how recent your experience was (e.g., what you had for lunch yesterday is easier to recall than what you had for lunch last week, last month, or last year). Also contributing to your ability to recall is the significance you give to certain events in your life. For example, you're more likely to remember your 16th birthday more than your 15th because it may have been perceived to be more important. Or you may recall with greater detail the birth of a child versus the day you got your first pet. For some, the day you got your pet may be a highly significant event. In simplistic terms, the regular and routine (nonspecial) events of daily life are stored in files that are not given the same prominence; thus they are not as readily recallable. It's those collectives of regular and routine events that I want to focus on for a moment.

Memories of routine experiences

Recall that many of your life's experiences are stored in various parts of your brain in a way you do not control. Under low stress conditions you are able to think things through in rational and logical ways. You can take the time to apply the traditional decision-making process discussed earlier; however, under stress (especially when time is compressed) it can be very difficult to process information and evaluate your options the same way. There simply may not be enough time. Fortunately, the intuitive process works well under stress. It should. Your brain has millions of years of practice in refining its intuitive skill set.

When you are under stress your brain searches through collections of experiences, looking for the solution to the problem. Fortunately it conducts these searches very fast, which is helpful in time-compressed emergency scene decision-making environments.

By comparison, think about the work environment of a commercial airline pilot. As long as the plane is flying properly, there is only a moderate amount of stress present. However, when something catastrophic goes wrong with the jet and the plane is rapidly losing altitude (while traveling at 500 mph) it's an understatement to say the stress level on the flight deck is high. The pilot may have only a few seconds or minutes to assess the problem and devise a course of action to prevent an otherwise inevitable bad outcome. Think of the quick actions needed by Captain Sullenburger when he ditched an Airbus A320 in the Hudson River on January 15, 2009, sparing the lives of 155 passengers and crew members.

When your brain is searching your stored images, it's looking for previous experiences that may help you solve your problem. If your brain is able to locate a similar experience (keeping in mind it doesn't have to be an *exact* match) it results in an intuitive gut feeling that indicates your brain has made a pattern match. Because these searches and matches happen at the subconscious level of the brain, the decision makers may not understand how they know what to do, they just know, intuitively, it's the right thing to do.

When Gary Klein's research group rode along with the expert fireground commanders in Cleveland, Ohio, they asked the commanders how they knew what to do. The common answer was, "*I don't know how I knew . . . I just knew*."

This unconscious knowledge possessed by those commanders—and by you—is known as *tacit knowledge*. As tacit knowledge resides in the subconscious, it's very likely you can't explain how or where you acquired it. One of the reasons you may not be able explain how you acquired it is that experiences (often in the form of stored images in the subconscious brain) may have been placed there without your awareness. In other words, your brain stores images of experiences you want to remember as well as images of experiences you don't even think about remembering. Recall, your brain stores nearly *everything*. Unfortunately, it also stores some things you might rather forget (like the image of that snazzy mullet you sported in high school).

On the flip side, if the decision maker has no previous experience or training related to the situation being faced, the brain may search the image files for the solution and come up empty. No match! When this happens, you're not going to get an intuitive gut feeling that tells you what to do. Some think the *absence* of a gut feeling is a good thing, but it can be foretelling that the decision maker may be essentially clueless about what to do because there are no stored experiences to draw upon.

This is why kids make so many mistakes; they lack experience. It's also why you, a parent, spend so much time trying to impart your wisdom to them, sharing your experiences with them, helping them to learn from *your* mistakes. This can be very helpful, even if you think they're not listening or heeding your advice.

This is also why novice responders are prone to making mistakes under stress. They don't have enough stored experiences to be able to effectively use an intuitive decision-making process under stress. Under stress, the novice is likely to have no pattern match; therefore, whether it's your kids or a novice first responder, you see their deer-in-the-headlights look. If you've ever observed it, you know what I'm talking about.

I stated earlier that when your brain is searching the images for a solution, the images (experiences) your brain identifies as the solution it is entirely likely you'll not have an exact match to some previous experience. Think about it—an *exact* match would be incredibly rare. (Personally, I can't think of two identical emergency calls I've responded to in my 30 years on the job.) Fortunately, your brain is pretty good about searching and combining together pieces of various experiences to help solve the problem. Whether it's from one experience or a variety of experiences combined, when your brain identifies the solution, it creates a *pattern match*. The pattern match is what causes the intuitive gut feeling.

While an intuitive gut feeling can guide decision making, the brain is not infallible; therefore, it is important to verify your gut feelings with facts that validate your intuition. A decision maker can get a pattern match that is incorrect, and the solution that feels intuitively correct may actually be wrong. It's important to seek out facts that validate the decision as a good one.

It's important to pause and issue a warning here. When decision makers assess a situation, capture clues and cues, form a pattern match, and get an intuitive gut feeling, it builds confidence. As the decision makers search for additional facts to validate intuition, it can cause them to look for the clues and cues that only affirm their intuition. In the process, clues and cues that refute their intuition may be dismissed or excused away. This phenomenon is known as *confirmation bias*.

Here's an experiment that demonstrates how your brain searches through your life's experiences in search of a pattern match, resulting in an intuitive gut feeling and the search for facts to validate the feeling:

You receive a 1,000-piece jigsaw puzzle in its box. The box is opened and the pieces are spilled out on a table. The researcher takes great care to ensure you don't get to see the picture on the lid of the box because it would make it too easy for you to figure out what the end result of the puzzle is supposed to look like. Relating this to emergency responses, it makes sense to do this because you never have a complete picture of the outcome when you arrive at an emergency situation, so you don't get to see the end result here either.

The likely first step for most participants is to turn all of the puzzle pieces upright and spread the pieces out on the table. This strategy serves two purposes. It allows the participants to see the colors, textures, and shapes of all the pieces. Think of each piece of the puzzle as a needed *clue* to solve the problem. Spreading the pieces out provides a way to scan all the clues and cues. This is critically important because some of the most important clues may be on the bottom of the pile, and if the participants don't spread the pieces out, they'll never know what they're missing.

The same principle can be applied to an emergency scene size-up. Conducting a complete size-up (i.e., looking at all sides of the incident) allows you to capture all the critical clues and cues. If the size-up is incomplete, you miss something that might be critical information.

The next likely step the puzzle assemblers take is to conduct a rapid scan of all the puzzle pieces. This allows them to capture small snippets of clues contained on each puzzle piece and make some assumptions about the picture's general characteristics and theme. Captured clues and cues might be parts of trees, water, sky, cars, buildings, people, words, animals, and so on. It would be nearly impossible to assemble all 1,000 pieces mentally (unless you possessed the prodigious memory abilities of Kim Peak, the man who inspired the movie *Rain Man*).

The next likely step is to locate the four corners. Why? Because the participant knows, with confidence, the corners are clues that contain a piece of *certain* information. The corners are great starting points from which the puzzle can be assembled because it can be deducted they go in one of four distinct places. Until some additional puzzle pieces are assembled, however, it's nearly impossible to know where any of the other 996 pieces go. Assume one of the four corners is blue. The participant doesn't have much to go on with that clue—it's blue and it's a corner piece. Further, assume the initial rapid scan of the pieces indicated the assembled picture would be an outdoor scene. So, with that assumption, what could a blue puzzle piece represent? Logic coupled with life experience (and perhaps some experience with jigsaw puzzles) might lead to the assumption a blue piece would be sky or water. Again, drawing on experience, it could be further assumed that if the blue piece is sky, it will be one of the top corners of the puzzle; if it's water, it will be one of the of bottom corners.

The next logical step would be to look for other pieces that fit together with the blue piece. Fortunately, experience with puzzles enables the participant to quickly sift through and eliminate (at least for now) all the pieces that do not have flat edges and are not blue. So, in essence, they are searching for the clues that *fit* with the blue corner piece. As multiple pieces are fit together, the participant is, essentially, forming a *pattern match*. The pieces that fit together provide additional information about the problem. Now, in addition to seeing just a single plain blue puzzle piece (which is *assumed* to be sky), the participant is able to identify pieces that have clouds and birds. The clouds and birds are *facts* that validate assumptions, and the participant can confidently deduct those pieces represent sky (see figure 3–10).

This is a very simplified example of tacit knowledge—how your brain searches through stored images of life experiences (the puzzle pieces) and fits the experiences together to form a pattern match that yields an intuitive gut feeling about what to do (or not to do). What makes the knowledge *tacit* is where the knowledge is stored and how it is used—subconsciously. Tacit knowledge allows you to form subconscious pattern matches without awareness.

Fig. 3–10. Assembling the pieces of a jigsaw puzzle is a lot like how the brain gathers and processes information in the decision-making process. Clues and cues (puzzle pieces) are captured and processed (put together).

There are multiple documented cases where people who have had near-death experiences reported seeing their lives flash before their eyes. Maybe you've heard of this happening, or perhaps you've been unfortunate enough to have a personal near-death experience and had it happen to you. The reports typically involve people who suddenly find themselves in a sudden and unexpectedly grave situation (e.g., in a vehicle running off the roadway and careening down a steep embankment at a high rate of speed). In the few seconds it takes for this event to occur, they report having visual images of their lives flashing before their eyes.

While having a discussion on the topic of decision making under stress with a group of firefighters, I asked how many had experienced near-death events and had their lives flash before their eyes. It was a fairly large group, and at least six hands went up. I asked one gentleman to share his experience with the group, and he shared an amazing story.

"I was a gunner on a helicopter in Vietnam in 1967. We were flying just above the trees when our helicopter was struck with a rocket-propelled grenade. The rotor seized up, and we were on a free-fall to the ground.

We were about 1,500 feet in the air when we got hit. As we were falling, I saw images of things from my life."

At this point, I prompted him to share the details about the images he saw.

"I saw an image of me at my high school graduation. I saw an image of me driving the first car I owned when I was 17. I saw an image of me riding a bicycle down a street in Germany when I was a kid. I saw an image of my first child being born. I saw an image from my wedding day. I saw an image of my grandmother in her casket at her funeral."

I asked how long he thought it took for the helicopter to fall to the ground.

"It was only a few seconds."

I was intrigued by the image of his grandmother's funeral, so I asked him where in the sequence that image occurred.

"That was the last image I saw before we hit the tree tops."

The gentleman further explained that these images weren't just in his head—they were what he was *actually* seeing . . . and doing! Everything in his peripheral vision went black, everything went silent, and these images were what his eyes saw as if he were actually in these locations and actually doing these activities. He said that they weren't images (still shots); they were movies playing out in real time, and he was the lead actor in each scene.

I asked him if he'd ever given much thought to what was happening to him that day and what it all meant. What he said next, if it were not a near tragedy, might have been quite funny. He said he did not tell anyone what happened because he was afraid they'd think he was nuts. Indeed it does seem odd that in the few seconds it took for his helicopter to crash he would see (and be part of) all those scenes—scenes that didn't really seem to have any connection to what was happening to him. When I shared with him an explanation of what was happening, I could see the relief on his face, as if to say, "Thank you for helping me to understand I'm not crazy."

What was happening is he was under stress. His brain was searching through his stored life experiences, trying desperately to find a pattern match that would give him the intuitive gut feeling about how to survive his crashing helicopter. Unfortunately, he hadn't experienced anything even close to a helicopter crash; thus, there was no pattern to match. No tacit knowledge. No gut feeling to tell him what to do. But what did happen was while his brain was searching through its stored visual images of his collective life's experiences, several of his life experiences rose from the subconscious level of his brain to the conscious level. And before him were actual relived experiences in real-time replay. Science hasn't been able to offer any plausible explanations as to why this happens for some people and not others, or which images are elevated to consciousness. Some images hold great importance (e.g., a wedding day or a child's birth) whereas others appear to be completely innocuous (riding a bicycle or riding a thrill ride at an amusement park). Nonetheless, it's a fascinating phenomenon that has been documented extensively.

I think it was quite ominous that the last scene this soldier lived out before his helicopter hit the ground was his grandmother's funeral. It's purely conjecture on my part, but this may have been his brain's way of saying it searched all the experiences on file and the only solution—the only way out of this stressful predicament he was in—was death.

The Role of Emotions in Decision Making

I want to go back to something I stated earlier in the chapter about the role of emotions in decision making. Rational, thinking people, supported by the findings of some research, believe there's no place for emotions in decision making—emotions should be set aside, and the decision should be made purely on rational, analytical facts. As many of us know, it's not that easy to control our emotions. Chances are some of the most important decisions a person will make in their life are heavily influenced by their emotions. Take, for example, choosing a life mate. Setting aside emotions in the decision would yield a purely rational, analytic evaluation of the lucky candidate. What are the person's personal hygiene habits? How compatible are you? Again, not based on how the person makes you feel—emotions are out. Base it on an analytical (1–10 scale) of measureable criteria. Remember to set minimum thresholds for what is acceptable and try your best not to let irrational, emotional things, like *love*, get in the way of making your good decision. In fact, ask someone what they love about the most important people in their lives and among the responses you're likely to get will be: *The way they make me feel when I'm around them*. But an advocate of purely rational, analytical decision making, would say there's no place for *feelings*.

One of the best books I've read on this subject is *Descartes' Error: Emotion, Reason, and the Human Brain* by Antonio R. Damasio. In the book, Damasio chronicles the studies of a man named Phineas Gage who suffered damage (from an injury) to the prefrontal cortex of his brain, which is the portion of the brain that controls his emotions. Gage was able to live a reasonably normal life, meaning he had no physiological deficit that affected his speech, hearing, vision, or motor skills. However, he lost several critical functions, including problem solving, maintaining attention, and the ability to control emotional impulses. This damage to the emotional control center of his brain also left him with an inability to make good choices and, to some respect, crippled his ability to make cogent, timely decisions among alternatives. For example, a discussion between Gage and his doctor about setting an appointment might take more than an hour because Gage could not decide on a good date or time. When he did make decisions they were often impulsive and inconsistent with his pre-accident demeanor.

In another study, a subject was engaged in a rigged card game where the cards being dealt followed a set pattern (see figure 3–11). The participants with normal prefrontal cortexes were able to pick up on the patterns, use intuition to help them anticipate the future cards, and avoid placing bets when they sensed the next card dealt was going to be bad (remember, the cards followed a pattern). The participants with damaged prefrontal cortexes were also able to identify a pattern within the cards, but they were unable to use intuition to anticipate when the bad cards were coming. Thus, they would continue to bet money even when the bad cards were inevitably going to be dealt. The lack of emotion essentially left a blind spot in their decision making.

Fig. 3–11. When the brain picks up on patterns of information it can predict future outcomes based on those established patterns. The next card dealt would be . . .?

Sliced and diced

Throughout this chapter I have been referring to how the brain stores lived experiences as images or scenes and stated, briefly, that it's a little more complex than that. Some research has shown that the brain does not store complete experiences as packaged images or scenes. Rather, the brain stores the various stimuli in separate regions of the brain. The things you see are stored one place. The things you hear are stored somewhere else. I recall hearing it described as like your experience being put into a blender with the lid off and pieces of the experiences are splattered all over your brain. Then, in a most mysterious way, your brain reassembles the experiences into coherent memories.

Under stress, as you gather the clues and cues in your current situation, you cannot consciously know whether it will be written words, spoken words, other sounds, images, colors, textures, aromas, odors, or what combination stimuli that will trigger a pattern match for you because, as you recall, it's all happening subconsciously. On a number of occasions I have encountered someone I know in a store or a restaurant. The chance meeting came unexpectedly and although I recognized the person as someone familiar to me, I could not immediately place a name with the face or even recall how I knew the person. After I walk away and give it some thought, I remember who the person is. Because so many of my professional associates wear uniforms, I am used to seeing them in uniform. In fact, part of the pattern that my brain matches (subconsciously) when I see them is their uniform *and* their faces—together. When one of those clues is missing (in this case, the uniform because the person is wearing civilian clothes), I struggle to make sense of who that person is. Eventually it comes to me, but not right away.

A similar thing can happen at an emergency scene. If you are used to regularly seeing a grouping of cues and clues that, together, hold some meaning for you and then one or more of those cues or clues are missing, you may struggle to make sense of what's happening. Your brain is efficient at taking pieces of a puzzle and putting them together correctly, resulting in something meaningful. In other words, an individual puzzle piece may not mean much until it is coupled with five or six other pieces. Then, and only then, is the meaning significant. But individually, that one clue or cue held little meaning for you. Much like you seek jigsaw puzzle pieces that fit together and connect them, your brain does the same thing. This process is called *chunking*.

To demonstrate how effective your brain is at chunking information, take a moment to read the following paragraph out loud.

> Aoccdrnig to rscheearch cnoedtcud at Cmabrigde Uinervtisy, it deosn't mttaer in waht oredr the ltteers are in a word. The olny iprmoetnt tihng is taht the frist and lsat ltteer are in the rghit pclae. The rset can be a toatl mses and you can sitll raed it wouthit a porbelm. Tihs is bcuseae yuor biarn deos not raed ervey lteter by istlef, but it chknus the ltters together and fmors the wrod as a wlohe.

Could you read the paragraph without a problem? Much to the chagrin of elementary school teachers who harp on the need for perfect spelling, most people can read it with ease, yet they don't really know how they can do it. The explanation lies in understanding that the letters of the words are chunked together in the brain and don't have to be in the proper order. The brain does need some facts to prime the recognition.

In the case of the example above, the facts are the correct location of the first and last letters.

I once conducted an impromptu experiment where I put the paragraph with misspelled words up on a screen and asked someone to read it. A participant expeditiously read the entire first line without realizing any of the words were misspelled. Then she abruptly stopped reading, paused, and said in a surprising voice: “You know you misspelled the word ‘university’?” The entire audience erupted in laughter, realizing something the participant had not—she read four misspelled words correctly without even realizing they were misspelled. After acknowledging my spelling error, I encouraged her to read on, and she did, reading the rest of the paragraph flawlessly.

I have seen, heard, and personally experienced the power of tacit knowledge driving decision making on numerous occasions. One call that comes to mind as a salient example was an incident I responded to where a fire was reported in a manufacturing facility. On my arrival there was smoke showing from the rear of this single-story 100-foot-by-400-foot warehouse-style building. I noted the location of the municipal fire hydrant on the boulevard, roughly 1,100 feet from the rear of the building. The access to the back of the building was a roadway that ran along the right side of the building. On my way in I noticed a private yard hydrant about half the distance to the back of the building. As I pulled around to the rear of the building I could see smoke was coming from an open garage door. I instructed the first-arriving engine to connect to the municipal hydrant at the street (even though I had noted the yard hydrant that was less than half the distance away). As the engine stopped to connect to the municipal hydrant, employees from the company were screaming at the company officer that there was a closer hydrant down the driveway. The company officer called me on the radio and relayed what the employees were saying. I instructed the company officer, again, to connect to the street hydrant and lay in. The engine laid out its entire bed of hose and came up short of reaching the rear of the building. This left the engine crew both confused and angry with me. I didn’t make the employees of the company very happy with me, either, and several came up to me immediately to express their displeasure. Thankfully our police department was able to quickly usher them away from the command vehicle so I could concentrate on the more important task of putting the fire out safely.

But it did raise a valid question about that decision. *Why would I do that?* I had to order the second-in engine to complete the hose lay while the first engine attacked the fire with tank water. On their way in, the third-in engine officer decided it would be a good idea to secure a second water supply, and they stopped to connect to the yard hydrant I'd ordered the first engine to pass by. From where I had established command, I could not see this hydrant so I was unaware the third-in engine had stopped to connect to it. Then I received a radio transmission from the third-in engine officer informing me that they had stopped to connect to the yard hydrant, but the hydrant was non-functioning.

At that moment, I went from being viewed as the dope that made the companies lay more than twice the hose needed and delaying the establishment of a patent water supply for the attack engine to being Nostradamus-like in my prediction that the yard hydrant would not work. When the incident was over, the officer on the first-in engine came up to me and asked, "*How did you know? How did you know that yard hydrant wasn't going to work?*" I wish I could have given him an answer that made me look smarter than I really am. The fact was, I didn't really know why I made that decision. Something inside me told me the yard hydrant was a bad bet.

At the time I gave the order I had no idea the first-in engine was going to run out of hose in the process of laying in to the scene. When that happened, I started to doubt myself and my order to abstain from connecting to the closer hydrant. Part of my doubt came because I didn't have a real good reason for the decision I made. I relied on intuition—I trusted my gut feeling—and made a decision that I would have struggled to defend. When I told the officer the reason I made that decision was I had a gut feeling the yard hydrant wasn't going to work, I don't think he believed me. As I left the scene, I stopped to look at the yard hydrant. I noted on this slower, more methodical inspection, that the hydrant top was rusted and the presence of weeds around the hydrant gave me the indication it had not been maintained. Was that enough to prime my intuition—a little rust and a few weeds (see figure 3–12)? Apparently so.

Fig. 3–12. Clues and cues—such as tall grass and the presence of rust—may be enough to trigger intuition and cause you to avoid connecting to a hydrant that appears to have not been maintained.

As I have spoken with hundreds of firefighters about decision making, they have shared with me countless examples of situations they have found themselves in where they describe a *gut feeling* that saved their lives. They couldn't explain how they knew; they just knew. I've had occasions where responders have shared with me that they experienced gut feelings that bad things were going to happen and they ignored those feelings . . . dismissing them because they could not readily find a conscious, rational explanation for feeling that way. Unfortunately, a number of those stories ended with bad things happening, including first responder casualties.

I am spending a lot of time on this topic because I want you to understand how important it is to understand intuition and the role it can play in your decision making. Those gut feelings are telling you something. You have to believe that when you get that gut feeling, it means something. Don't dismiss it. However, I am also going to caution you. You can get gut feelings that are wrong!

Good, But Not Perfect

Although your brain does a wonderful job of searching your collective life experiences for the solution to problems when you're under stress, you can have a pattern match that is wrong. To help you overcome this, when you get that intuitive gut feeling, remember to look for some facts to validate your feelings.

Equally, look for facts that *refute* your gut feelings. This is an important point that you should not take lightly. When you think you have something figured out in your mind, it's natural to look for data and information that validates you are right. You do this all the time when you deal with people. If you have a good idea, you seek out the people who you believe will be supportive and encouraging of your good idea. Likewise, you may avoid talking to those who you suspect are going to be negative. This can be problematic because the negative feedback may actually be more valuable in the development of your idea than the supportive and encouraging feedback. When you are trying to validate your gut feelings, seek information that both validates your feelings as well as refutes them. Both provide valuable clues and cues.

Another word of caution is in order. Some commanders may struggle with knowing how much factual information to gather to confirm they are making a good decision and consequently delay their decision or make no decision at all. This *analysis paralysis* can be dangerous. If you wait until you have 100 percent of the information needed to make an emergency scene decision, you're probably going to have a bad outcome because as soon as you think you have 100 percent quality and reliable information, a large portion of that information has changed.

First responders are action oriented. If a commander is unable or unwilling to make a decision because of incomplete information, there's a risk the responders may grow impatient and engage in independent goal setting—*freelancing*. I'm not condoning independent action; rather, I'm simply making an observation about what can happen when timely decisions are not communicated. A good rule of thumb, depending on the emergency, might be to gather 30 to 70 percent good, solid information to validate your intuition, and then make a decision and go with it.

Four Essential Requirements for Recognition-Primed Decisions (RPD)

To be able to make good recognition-primed decisions, a decision maker needs the following four essential skills.

RPD requirement #1: Situational awareness

The decision maker must develop and maintain strong situational awareness. A simple and often-used definition of situational awareness is *paying attention*. However, as is discussed throughout this book, paying attention can be much harder than it appears. A review of near-miss and line-of-duty death reports may lead to the assumption that those events could have been avoided if the decision makers were simply paying attention to what was going on around them. This is not to say the decision makers were purposefully *not* paying attention.

As we explore in the chapters ahead, there are dozens of formidable barriers that challenge decision makers' situational awareness and numerous ways situational awareness can be eroded. Some of them are obvious and overt. Others are subtle and covert, robbing decision makers of their situational awareness in ways they may not even realize it is happening.

RPD requirement #2: Tacit knowledge

To make recognition-primed decisions, it's essential for the decision maker to have tacit knowledge—that unconscious knowledge experts possess. This comes from years of collective training and experience and resides mostly in subconscious memory stores. Everyone possesses tacit knowledge to some degree on some topic. The tacit knowledge essential for public safety decision making comes from extensive development of memories of training and emergency event responses.

RPD requirement #3: Mental modeling

To make good recognition-primed decisions, the decision maker must have the ability to develop mental models. Mental modeling occurs when the decision maker takes what is happening at the current moment on the emergency scene and, drawing upon training and experience, visualizes a plan of action. These models display the future results of their decisions. Simply stated, mental modeling is making predictions of the future. The reason this is so important is accurate mental modeling can foretell potential bad things well before they actually happen. When a decision maker develops a mental model that results in a bad intuitive gut feeling, that's a strong indication there is a significant cause for concern. Mental modeling can also help a decision maker set some benchmarks for progress. As the decision maker plays the proverbial *mental movie* forward, certain events are naturally anticipated. Those events become benchmarks.

For example, on arrival at a room-and-contents residential-dwelling fire, the decision maker plays the mental movie into the future and imagines assigning the first-in company to attack the fire with a hand line. As the decision maker sizes up the structure and the crew's abilities, he or she imagines how long it might take for the crew to traverse the rooms and obstacles in the house to reach the fire. This helps the decision maker establish certain expectations about how long it should take before seeing the expected results. The decision maker also imagines what those expected results might look like after the crew reaches the fire and applies water (e.g., the transition of black smoke to a whole lot of white smoke). Mental modeling can help a decision maker anticipate bad things that could happen before resources are committed to an action that may be difficult to reverse.

One way decision makers can get into trouble is failing to appreciate how important mental modeling is and, subsequently, failing to take the time to play their mental movie forward and think about what's going to be happening at the emergency scene 10 or 20 minutes into the future. A commander who thinks only in the current moment is asking for trouble. If you think about it, this is true in any environment where a human being is working in an environment with rapidly changing conditions.

A good example that demonstrates the importance of mental modeling and predicting future outcomes is the revealed as I share the first and only time I ever rode a Jet Ski in the ocean. I was a *novice* Jet Ski rider (and it showed). The ocean was kind of choppy that day; the waves looked 30 feet high to me. At first I went slowly, and things went well. As I got my confidence up (that is, after 10 minutes of developing "expertise"), I started opening up the throttle and going faster. After a couple nasty spills I quickly realized the trick to staying upright while going fast, and living to tell about it, was to look beyond the wave right in front of me to the waves 20 or 30 feet ahead so I could anticipate what I would be facing as I came off the wave I was currently riding. Before I developed this expertise, I was not anticipating the future waves. Rather, I was riding only the current wave. I was dealing with the problem I was facing in the current moment in time and not worrying about the one that was coming next. Getting caught off guard by not looking out ahead of the current moment had less-than-enjoyable consequences (ouch!).

It's not difficult for a decision maker to fall into the trap of failing to develop mental models of future events. If the incident is rapidly changing and there are lots of clues and cues to capture and comprehend, the decision maker may struggle just to make sense of what is happening in the current moment, let alone possess the additional cognitive capacity to mentally get ahead of the incident and think about what is going to happen in the next 5 or 10 minutes. The best mental modelers possess extensive expertise (acquired through training and experience) that aids in the rapid development of accurate predictions.

However, training and experience do not, in all cases, ensure the development of a strong mental model. Here's an example to demonstrate the point.

Imagine a first responder with years of experience and extensive training who is very proficient performing *interior* structural firefighting operations. Being promoted to a command-level position (i.e., the person in charge on the *exterior* of the structure), the person may have a high level of confidence about his or her expert knowledge based on years of experience. The problem is the clues and cues captured and processed in the command role can be very different from the clues and cues captured and processed in the operational role. The decision maker has years of training and experience, but is a *novice* commander. Experienced commanders know that in addition to the interior and the exterior of a structure fire looking, sounding, and feeling very differently, it's also a distinctly different decision-making environment.

RPD requirement #4: Self-confidence

The fourth and final requirement for making good recognition primed decisions is self-efficacy, or the more commonly used term, self-confidence. A decision maker must be willing to trust his or her intuition, and that trust comes in part from understanding the origins of intuition. (See figure 3–13.)

Fig. 3–13. Experienced commanders have all four components required to use RPD: situational awareness, tacit knowledge, mental modeling, and self-confidence.

> When intuition leads to the feeling that *something's not right—I can't explain why, but something isn't right about this situation*, the decision maker must be willing to trust the mental pattern match that has triggered this intuition.

Commanders who do not understand the origin of intuition or who lack self-confidence in their intuition are very likely not to trust their gut feelings. The presence of a gut feeling that leads to a sense that something is not right should cause the decision maker to pause and start asking hard questions about the meaning of those feelings. A decision maker who tries to make rational sense out of gut feelings can quickly dismiss the feeling because there may be no overt, measurable evidence that anything is wrong. Be very careful about allowing rational, logical thinking to overrule the concerns sparked by intuition.

CHAPTER 4

SITUATIONAL AWARENESS

Recall from an earlier discussion the simple definition of situational awareness—paying attention to what is happening around you. This certainly is part of situational awareness, but not all of it. It's plausible that a decision maker who is *not* paying attention would find it challenging to be situationally aware. For example, it would be difficult to be situationally aware if the decision maker was asleep or unconscious. That seems obvious.

What may be far less obvious is the impact of chronic and acute fatigue on situational awareness. And another less-than-obvious impact on one's ability to pay attention is multitasking. For example, performing a physical task while commanding in a dynamically changing environment can impact the ability to process information and pay attention to changes in the environment. Here's an experiment you can do that demonstrates the challenges (and consequences) of multitasking. Take someone out on a brisk walk. After you are up to full stride, surprise your companion by asking the following question: "Don't guess your answer . . . What is 64 times 17?" It is almost a guarantee the person will stop walking immediately because devising the answer requires total brain power. It will not be possible to walk and perform the complex mental calculation simultaneously. If your companion does continue to walk (maybe out of courtesy or sense of obligation not to hold up the walk), it will take much longer to come up with the right answer.

Developing and maintaining situational awareness is not simply a matter of being conscious and showing up. If it were that easy, there would be no need for the millions of dollars being spent on research to improve high-risk decision making, and there'd be no audience for this book. Likewise, the number of workplace accidents and injuries (far beyond emergency services) would be vastly reduced. The bad news is that developing and maintaining situational awareness is far more complex than many public safety providers realize, and this lack of understanding of the complexities causes some to dismiss or discount its importance. The good news is that the more a decision maker knows about how situational awareness is developed, maintained, lost, and regained, the easier it is to learn how to deal with the challenges of making quality decisions under stress.

In my quest to understand situational awareness on a deeper level, a couple of things became apparent. I found that the amount of scientific research conducted on situational awareness in public safety domains is woefully inadequate. There is no shortage of articles written by first responders on the topic. Some include suggestions for how to develop and improve situational awareness based on the authors' personal experiences, and there's nothing wrong with that. In fact, much can be learned from those articles, especially when they contain examples and advice about how decision makers developed and maintained their situational awareness in high-stress, high-consequence emergency situations. The problem is, some of these authors may not understand the contributing factors to their strong or eroded situational awareness. In part, you could blame it on tacit knowledge. They know what they did (or didn't do), but they may not understand the factors that contributed to developing or eroding their situational awareness, which can leave them falling short on offering good, sound advice about the lessons learned (figure 4–1). They may also be experts at pointing out the mistakes captured in video and audio vignettes of incidents gone bad, but they don't really understand the deeper, more complex issues related to the decision maker's situational awareness.

As often is the case in popular trade journals, many articles contain anecdotal evidence and are rarely based on scientific research of the subject. In one sense, that's a good thing because there are many public

Fig. 4–1. The lessons of situational awareness can be learned through understanding the science of decision making as well as through anecdotes from lived experiences.

safety professionals who are turned off by scientific studies and they would probably not enjoy reading research studies.

The information contained in this section of the book is based on my original research. *Original* means the researcher designs the project with the goal of discovering *new* information—something that has not previously been discovered. This is the opportunity that was laid at my feet when I chose to complete my doctoral research on first responder decision making and situational awareness.

One of the first things I needed to do was to pick a working definition for *situational awareness*. As it turns out, I found seven different definitions for situational awareness in my review of the literature. The definition that I believe best fits public safety providers was created by a brilliant researcher, Dr. Mica Endsley, whose company, SA Technologies (*SA* standing for situational awareness), works extensively in the area of situational awareness research, design, training, and measurement. Many of Dr. Endsley's articles are available on the company's website: www.SATechnologies.com. The Endsley group has completed extensive research in aviation, air traffic control, space exploration, maintenance, power systems, transportation, medicine, and the military.

> Endsley describes situational awareness as "*Being aware of what is happening around you and understanding what that information means to you now and in the future.*"

Dr. Endsley postulates the three levels of situational awareness.

Level 1 Situational Awareness

The first level of situational awareness occurs when the decision maker captures the cues and the clues in the current situation. These clues and cues can be captured both consciously (overt and purposeful use of the senses) and subconsciously (captured without realizing it). In other words, when a decision maker assesses an emergency scene, the eyes and ears may focus on one or more pieces of data, but the brain is taking in and processing much more than the data points the eyes and ears are focused on.

Level 1 situational awareness is key to a decision maker's ability to conduct an effective size-up. If a decision maker expects to have any chance of assembling a coherent understanding of what is happening, he or she must be able to capture clues and cues (using all their senses). Think of clues and cues as puzzle pieces and Level 1 situational awareness as gathering the puzzle pieces.

There are many well-documented examples in near-miss and line-of-duty death reports where decision makers failed, for whatever reason, to conduct a complete size-up. Many reasons have been suggested as to why this happens, and those reasons will be addressed throughout the remainder of this book. Stated simply, decision makers cannot develop strong Level 1 situational awareness if they fail to capture the cues and clues in the environment.

The process of forming Level 1 situational awareness likely starts well ahead of receiving the notification of an emergency call. In the conscious human, the senses are always, to some extent, scanning the environment—taking in and processing information. If the decision maker wakes up in the morning and looks out the bedroom window, the brain captures the weather conditions. Maybe the decision maker watches or listens to the morning news and forecast, and the brain captures and stores more information about upcoming events. Things are observed while driving to work are captured and stored. When dispatched to an emergency call, the brain captures more information. The dispatcher notifies the decision maker of the address and nature of the emergency. The decision maker may recall clues and cues previously captured and stored about that particular address, the neighborhood, construction of the homes, water supplies in the area, or typical traffic conditions on the intended route of travel for that time of day. Dispatch may also provide additional clues and cues about what is being reported. Endsley terms Level 1 situational awareness as the *perception* phase—capturing the information about what is happening. You can think of it as *paying attention*, and it's the beginning of situational awareness formation.

Level 2 Situational Awareness

The next level in the development of situational awareness is Level 2, which Endsley terms the *comprehension* phase. At this point, the decision maker takes the clues and cues captured in Level 1 and makes sense of them. This is where the decision maker figures out what is going on—deriving meaning from the combined clues and cues. Applying the jigsaw metaphor from chapter 3, Level 1 situational awareness is the equivalent of taking all the puzzle pieces out of the box, spreading them out on the table top, turning all the pieces face up, and taking a broad scan of all the colors, textures, and shapes. Level 2 is where the assembly of puzzle pieces begins. As a few of the pieces fit together, things may begin to make more sense. The puzzle assembler may start with a corner and look for another edge piece with a similar color and texture and a complementary shape. As the pieces fit together, an image begins to form that bears some meaning. This is what the brain does during Level 2 situational awareness. It takes the clues and cues from Level 1 and puts them together to mean something. The decision maker *comprehends* what is happening.

This is the point in the development of situational awareness where the decision maker might be thinking "*What does all this mean?*" All the bits and pieces and chunks of information have been gathered. Now they need to be assembled into meaning.

Much like my description of Level 1 situational awareness, the development of Level 2 may appear to be a simple task—look at clues and cues, and figure out what they mean. Unfortunately, it's not always an easy thing to do, especially if some of the clues and cues are unfamiliar to the decision maker or if the clues and cues do not fit the decision maker's beliefs about what he or she *should* be seeing in the current moment, regardless of what is *actually* being seen. As the barriers to situational awareness are discussed in detail in the coming chapters, it will become more apparent as to how challenging it can be to figure out what is happening, especially when the decision maker is under stress.

Level 3 Situational Awareness

The final level in the development of situational awareness is Level 3. Endsley's term for this level is *projection*. Here, decision makers make assumptions and predictions about future events. The clues and cues captured at Level 1 are comprehended into meaning at Level 2. Now comes the highest and perhaps the most complex level of situational awareness—being able to predict where the incident is headed next.

To make projections about the future, a decision maker must mentally get ahead of the current moment and think about what is going to happen at a future point in time (e.g., 5, 10, 20 minutes from now). This can be tough to do, and decision makers definitely become better at predicting future events as they gain experience.

All three levels of situational awareness are interdependent on each other. They can build on each other. For example, strong Level 2 situational awareness is predicated on capturing clues and cues. Likewise, it would be difficult to predict the future of an incident if the decision maker did not have a good understanding of what is happening at the current moment.

In summary, Level 1 situational awareness is using *perception* to capture the clues and cues in the environment. This happens both consciously and subconsciously. Level 2 situational awareness is using *comprehension* skills to figure out what the clues and cues mean—sorting them out and making sense of them. Level 3 situational awareness is making *projections* of what is going to happen next—foretelling future events.

On the scene of an emergency, developing and maintaining each level of situational awareness is not a one-time process; rather, it's continual. Because the incident is ever changing (and often rapidly changing), decision makers must continually use their senses to capture clues and cues, comprehend them into meaning, and adjust their predictions about the future accordingly. If all decision makers had to do at an emergency scene was to develop and maintain their situational awareness, this might be an easier task; however, experienced decision makers know there is so much that need to be done.

The compression of time can also challenge situational awareness, and stress can cause an inaccurate accounting of the passage of time. *Temporal distortion*—when stress causes time distortion—can make it difficult for a decision maker to effectively predict the future. Say, for example, the decision maker's mental model reveals a certain expectation 10 minutes into the incident. When stress causes time distortion, 20 minutes may pass, whereas for the decision maker, it may feel as though only 5 minutes has elapsed (figure 4–2).

Fig. 4–2. While operating under stressful conditions, time can become distorted or compressed, making it a critical situational awareness best practice to keep track of the passage of time.

Experienced commanders (even those unfamiliar with Endsley's work) know how important it is to perform each step and probably have been the victim of the consequences when they didn't. Yet, all decision makers have one fallible trait they must reckon with. They are human, and their brains have a limited capacity to process and understand information, especially in dynamic environments. The limited capacity to capture and comprehend information has nothing to do with intelligence. It has to do with how the brain processes information. Are there exceptions to the rule (norm)? Absolutely. Recall the average person's brain can capture, comprehend, and recall about seven (plus or minus two) pieces of unrelated information. Does that limitation apply to everyone? No. There are few things in the world of brain science that follow absolute rules. There are always exceptions—people with special gifts and talents.

Subject matter expertise (through training and experience) is one of the most effective ways to improve a decision maker's ability to develop, maintain, and regain situational awareness (and to effectively apply the recognition-primed decision-making process). A newly hired police officer, firefighter, or emergency medical technician cannot possess subject matter expertise after completing basic skills training and responding to just a few emergency calls.

The Making of an Expert

How much training and experience does it take for someone to be considered an *expert*? That is a researchable question, and it has been studied rather extensively in a variety of professionals including athletes, musicians, and physicians. The results of the research may leave you shocked—maybe even a little despondent.

> Research suggests it takes, on average, *ten years* of regular and routine acquisition of knowledge and skill practice to develop expert-level knowledge and performance if a student is in an environment of learning and/or practice two hours each day, five days each week.

If those research results for expertise were applied to public safety, there would be few who would qualify as *experts* (figure 4–3).

Most of the public safety providers I interact with are fairly seasoned public safety decision makers; however, on one occasion I was speaking to a department's medical director. He shared with me that he, on occasion, rides along with the fire department to observe their performance. I asked the physician if he had any fire training. It turns out he had no fire training and, although he had responded to a few fire calls, he responded mostly to medical calls. He was an incredibly intelligent person, but he had no fire training and very little experience. I asked the doctor if he were at a fire scene and observed a large volumes of black smoke emitting from the building that turned into large volumes of white puffy smoke, what that would mean. He said he had no idea. The firefighters we were with at the time were mildly entertained. They quickly explained to him that when water is applied to a fire it turns to steam, and that puffy white smoke is the steam, indicating the fire is being extinguished. He had no idea. Coincidentally, this conversation advanced his competency. He now knows what that would mean if he saw it.

Fig. 4–3. It takes, on average, 10 years of command experience to acquire the knowledge and skills needed to become an expert incident commander.

The Progression of Competency

Decision makers acquire their expertise in a four-step process of progression. These steps are explained in the following sections.

Unconscious incompetence

The first level of the competency progression is *unconscious incompetence.* People who are unconsciously incompetent are unaware (unconscious) about how little they know (incompetence). Stated another way, they lack so much training and experience that they don't even know what they don't know (figure 4–4). The cliché "ignorance is bliss" applies to the person who is unconsciously incompetent.

Fig. 4–4. Every first responder starts at the unconscious incompetent level of knowledge—they don't know what they don't know.

If an unconsciously incompetent person were on the scene of an emergency, and some significant yet subtle clue or cue presented itself, it is very likely this person would not even notice it. The person doesn't know enough to be looking for cues and clues. Everyone starts out being unconsciously incompetent—knowing nothing about a subject matter. Hopefully they don't spend much time at this stage of their competency development before they start acquiring knowledge and experience, especially as it relates to providing emergency services.

Conscious incompetence

The second stage of competency progression is *conscious incompetence*. At this stage, decision makers become aware of how much they don't know. People progress to this level of competency by taking classes and responding to calls. The acquisition of this new knowledge leads to an epiphany. A consciously incompetent person might think "*Holy cow! There's a lot I don't know!*" When that happens, that person has graduated to being consciously incompetent. It can be a scary thing when this happens, because decision makers who are consciously incompetent don't know enough about the ways they could get hurt or killed on the job—and they know it.

This step of competency progression can be a good thing. At least the decision makers now know enough to be cautious at emergency scenes. The decision maker can still get caught up in bad situations, because they may not necessarily know enough about how to avoid them. People who are consciously incompetent might respond to a call where something happens and say "*I saw it, but I didn't know what it meant.*" Their limited training and experience was enough for them to take notice of something happening at the incident scene, but they didn't know enough to fully understand what it meant.

Conscious competency

As decision makers gain more training and experience, say three to five years, they build a broader knowledge base and advance to the next level of competency development—*conscious competency*. As they reach this level, decision makers begin to become aware, at the conscious level, just how good they are becoming, and they are consciously aware of how knowledgeable they have become. They're competent decision makers, and they know it. At this stage of the competency development, decision makers are gaining (and displaying) confidence in their abilities. When they reach this level of competency development, they would respond to an incident, see something happen, and say "*I saw it . . . and I knew what it meant.*" Their training and experience pays off because they are able to readily recall the lessons they have learned to aid in their decision making.

Unconscious competency

As decision makers gain even more experience and obtain additional training, their competency develops to its highest level—*unconscious competency*. When you arrive here, you become so good at what you do, so experienced, knowledgeable, and well practiced, that knowing what to do in a high-stress, high-consequence situation seems to come naturally and automatically. Your decisions are aided by intuition, derived from tacit knowledge (the collection of your training and experience).

Reflect back to a time when you were watching a professional sporting event. It's likely the athletes were so well trained and well practiced at their sport that their performance seemed natural and automatic. This level of performance doesn't happen by accident. To the contrary, it comes from years of intensive training and experience that pays off when they are under pressure to perform. First responders who have developed their highest level of competency are able to process information automatically and effortlessly.

I'm reminded of a lesson I learned from Dr. Denis Waitley, a well-known psychologist, lecturer, and author of one of my favorite books, *The Psychology of Winning*. Among his many career accomplishments, Dr. Waitley served as an advisor to the 1984 U.S. Olympic Team and a young gymnast named Mary Lou Retton. While the coaches worked on developing the physical competencies of the athletes, Dr. Waitley worked on developing their psychological competencies. Waitley knew that behaviors practiced over and over again become psychologically engrained as habits, that under stress, would guide the athletes' performance. He also knew that if the athletes were not well practiced or if they lacked confidence in their abilities, they may start second guessing themselves, and that this could create stress and tension, and inhibit their performance.

Waitley knew that under stress, his well-prepared, well-practiced gymnasts could perform their routines without thinking about it at the conscious level. Their bodies knew what to do based on habits they developed through practice. When training is lacking, the confidence to trust tacit knowledge may also suffer. Waitley told his athletes to stop thinking about their routines—don't let your mind get in the way of what your body knows how to do. Let the body do what it has been trained to do. Mary Lou Retton had unconscious competency. All she had to do was have the confidence in herself to perform her routines without letting her mind get in the way. And she did. Her routines were flawless, and she won the gold!

Commanders who have developed unconscious competency can respond to an emergency, conduct a size-up, and know, intuitively, what things mean and what needs to be done. If asked, they might say "*something inside me told me that was an important piece of information*." Yet they may not be able to articulate what it was that triggered their reactions.

A good example of how a person develops competency can be illustrated with the task of driving an automobile. As newly licensed drivers, teenagers are unconsciously incompetent when it comes to the act of driving. Even though they may have observed others driving a vehicle hundreds of times, they are unaware of all the knowledge, skills, and abilities needed to operate a motor vehicle; however, adults who have mastered the art of operating a vehicle by practicing thousands of times are able to drive a vehicle without much conscious thought.

> Even though an experienced adult driver can operate a vehicle using mostly subconscious automatic scripted programming, it's still a good idea to pay attention while driving. This includes the need to avoid undue visual and audible stimulation and interruptions.

As drivers develop unconscious competency, they can capture and process clues and cues subconsciously, and their bodies can react to environmental stimuli without having to spend much time thinking about it.

I recall many times in my life where I drove somewhere and had no conscious recall of portions of the trip. The scary thing is (and a little embarrassing to admit) sometimes those experiences involved driving an emergency vehicle to a call (figure 4–5).

Fig. 4–5. As much as it may be unnerving to think about, an experienced apparatus operator drives to emergency scenes in part using subconscious programming. Novice apparatus operators cannot do the same thing because they do not have sufficient driving experiences stored in their memory. This makes the novice operator far more likely to miss critical clues and cues on the roadway.

Reflecting back, I had so many things on my mind that the act of driving was occurring at the subconscious level. Maybe it was just luck that prevented me from ever being in a serious accident, assisted some perhaps, by my unconscious competency for driving.

Think about all the things that happen when arriving at the scene of an emergency. Absent the experience that leads to unconscious competency, the decision maker has to spend precious time and a lot of mental energy taking in cues and clues and processing them into meaning at a conscious level. Because they have developed unconscious competency, experts are able to scan an incident scene and know, intuitively, what is going on and what needs to be done. Thus, an expert's limited conscious capacity is not consumed by the process of making sense of environmental clues and cues.

The application of competency development to driving may help to explain why young drivers are prone to have so many accidents. They have not developed their driving competency to a level where they can process cues and clues and know, intuitively, how to respond. They process information at the conscious level and, restricted by the brain's limited working memory capacity, they can easily become overloaded. When they are behind the wheel with something on their mind, or if they are carrying on a conversation with someone in the car, or if they are talking on their cell phone or texting, they're being set up to have an accident. They don't have enough training and experience to operate with unconscious competency, and their conscious (working memory) will be easily overloaded. It's a recipe for disaster (figure 4–6).

(Of course, the same consequences can befall experienced drivers who talk on cell phones or text while driving.)

Fig. 4–6. A young driver requires conscious awareness to operate a vehicle. Distractions, such as texting while driving, or even talking with someone else in the vehicle, can be enough to draw their attention away from the roadway.

Experts process and make sense of the clues and cues on a mostly subconscious level. Studies suggest the subconscious brain is able to capture and process cues and clues and alert decision makers to dangerous situations before they are even consciously aware of the danger. Researchers have conducted experiments where they have connected sensors to brain receptor regions that detect stress and danger. The test subjects were shown a movie and instructed to press a button when they sensed something scary was going to happen. Amazingly, the receptors of the brain were activated well in advance of the person being consciously aware that something happened. This is because the senses captured cues and clues and processed them into meaning at the subconscious level, well ahead of conscious awareness.

As noted earlier in this chapter, a critical component for being able to make effective recognition-primed decisions is developing and maintaining strong situational awareness. The next chapter provides an abbreviated overview of research conducted on decision making and situational awareness with first responders, followed by research findings that identify and discuss the barriers that challenge emergency scene decision makers' situational awareness. Many of these barriers can keep responders from seeing the bad things coming in time to change the outcome.

CHAPTER 5

SITUATIONAL AWARENESS RESEARCH

As part of my journey to complete my doctor of philosophy degree I was given the opportunity to conduct original research on an incredibly fascinating topic—how first responders make decisions in high-stress, high-consequence dynamically changing environments. The deeper I immersed myself into this topic the more amazed I became to learn the number of barriers that can impact a decision maker's situational awareness. Whether you are an expert decision maker or a novice in the process of developing your expertise, I hope you will find the information contained in the following chapters on situational awareness barriers thought provoking and informative.

When I started conducting my research, it quickly became evident to me how little *real* research had been conducted in the emergency services field on the topics of decision making and situational awareness. The research I conducted focused on the barriers that keep decision makers from seeing bad things coming in time to change the outcome. I am honored to share these barriers and provide some advice on how to overcome them. From my experience with training and working directly with thousands of first responders, I know if the decision makers could have seen the bad things coming, they would have made different decisions and possibly avoided the near miss or the catastrophic event.

The study I conducted used expert-level decision makers. The scientific rigor required for my study required me to narrowly focus my study; thus, fireground commanders were my subject group, and residential dwelling fire scenes were the operational environment. However, in the broad sense, the context of the study participants or their decision-making environment proved irrelevant to the recommendations for improving situational awareness. In other words, the barriers to situational awareness for high-stress, high-consequence decision makers are universal. This is an important point, because regardless of your profession, the findings apply.

Expert Decision Makers

As I had established previously in this book, recognition-primed decisions require expertise, and one element for making good decisions is situational awareness. It was an important requirement for the decision makers who participated in my study to be experts because novice commanders make decisions very differently than experts. My interest was focused on the challenges that *expert* decision makers face in developing and maintaining situational awareness at emergency scenes.

To ensure my participants would be experts, they had to meet a minimum threshold for their training and experience as commanders of residential dwelling fire incidents. Whereas there are some common challenges decision makers face at all emergency scenes, there are also some unique features about each type of emergency. For example, the situations that challenge a decision maker at a dwelling fire can be different from a high-rise fire and different from a hazardous materials incident. I chose residential dwelling fires because they account for the highest number of first-responder casualties.

The study participants had, on average, 20 years of emergency services experience. The average years of experience they served as an officer was 13, and they averaged 11 years of experience as command-level decision makers. The participants had supervised, on average, 105 residential dwelling fires. It can be said, confidently, they were an experienced group of decision makers (figure 5–1).

Fig. 5–1. Commanders who participated in Dr. Gasaway's research were required to meet minimum criteria for training and experience.

To understand the challenges a decision maker faces in developing and maintaining situational awareness, it may be helpful to understand the many and various ways a decision maker's situational awareness can be challenged while working in high-stress, high-consequence dynamically changing environments. Although it was disheartening to learn how little scientific research had been done on this topic within the emergency services domain, it was valuable to learn the topic had been studied extensively in other high-risk, high-consequence fields.

An evaluation of the existing research revealed many common threads for decision making among those working in similar high-risk, high-consequence domains. For example, in one study a flight attendant interrupted a pilot during the preflight checklist, causing the pilot to overlook an important preflight task (figure 5–2). In another study, a surgeon was distracted by the sound of a repetitive alarm sounding on a ventilator (figure 5–3). In both cases the results of these distractions and interruptions were measurable and catastrophic.

Fig. 5–2. Pilot completing preflight checklist

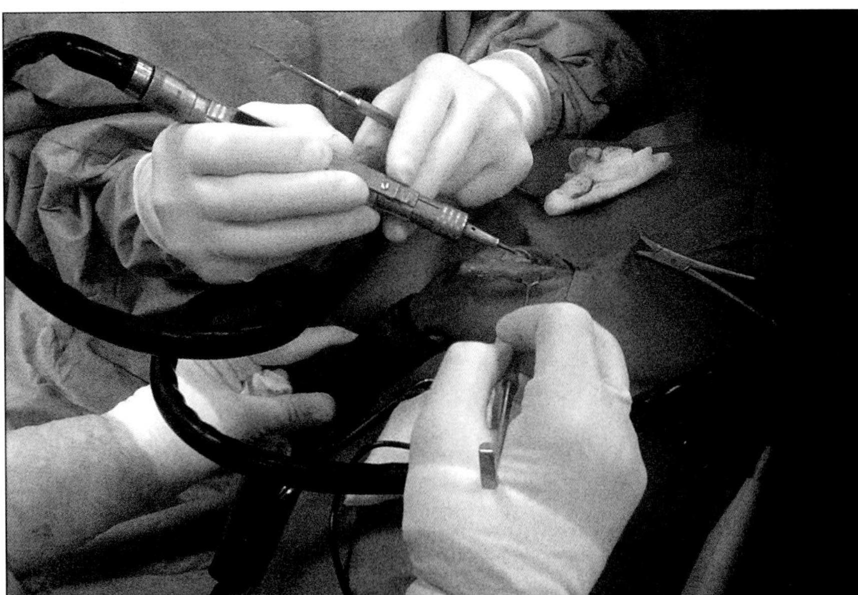

Fig. 5–3. Noises in an operating room, such as an alarm on a ventilator, may be enough to distract a surgeon.

While evaluating such studies, the underlying question I asked was, "*Could a public safety decision maker be challenged by the same barriers as pilots, surgeons, and military decision makers and, as a result, lose their situational awareness?*" If it were plausible, I listed the root cause as a *potential* barrier for public safety situational awareness. Over a six-month period I reviewed dozens of fascinating studies on the challenges decision makers face in commercial and military aviation, surgery, emergency rooms, law enforcement, battlefields, air traffic control, submarine command, and more—all decision environments with similar challenges to public safety decision makers: high-risk, high-consequence, dynamically changing environments, unknown cause and effect, and high ambiguity.

During a discussion on public safety situational awareness barriers, someone challenged me when I compared the decision-making environment of a commercial airline pilot to the decision-making environment of a first responder. From this gentleman's perspective, being a commercial airline pilot was nothing like being a first responder. "*With the technology on planes these days, the aircraft nearly flies itself,*" he retorted. I agreed with him completely. Modern commercial airlines are certainly high-tech and full of gadgetry designed to make sure the passengers make it safely to their destinations (and, as someone who flies a lot, I am thankful for that). But he missed the point I was trying to make. When all the systems are working as they are supposed to, a plane can (almost) fly itself. However, when something goes wrong, the decision-making environment changes quickly.

Take, for example, the double bird strike that brought down U.S. Airways flight 1549. Captain Chesley Sullenberger was pilot in command. Following the bird strike the plane was in a rapid descent due to loss of power in both engines. As you can image, that was a high-stress, high-consequence work environment for the flight crew. Unfortunately, the crew had little time to reference manuals and protocols. There was a lot at stake, including the potential for the entire crew and all passengers to lose their lives if his decision making was flawed. The training that pilots receive to prepare them for making such critical decisions under these conditions can provide valuable insights for first responders so long as we remain open minded to learning from how other high-risk, high-consequence professions train and develop expertise.

A Staggering Number of Barriers

When I set out to identify all the ways responders could have their situational awareness challenged I thought there might be upwards to 20 different barriers. Wow, was I wrong! The list of 20 grew to 40. Then 60. Then 80! By the time I was done compiling the list, I had 116 potential barriers that could challenge a decision maker's situational awareness. I recall at one point looking through the list and thinking to myself, "*With this many potential barriers, how could a decision maker ever develop and maintain good situational awareness at an emergency scene?*" It was rather discouraging.

Sort out the barriers

This overwhelming list would surely make the next task even more challenging for me as I would try to identify which barriers, among those 116, were most challenging for emergency scene decision makers. The method that I chose to vet the list is called a *multiple card sort technique*. Each barrier was listed on an index card, and a sample sentence was provided for context. For example, one card might say, "*Distractions: My situational awareness has been challenged by* distractions." (See figure 5–4.)

Each participant worked on the card sort exercise alone and in a controlled environment (e.g., a classroom, meeting room, or office). A board divided with three boxes was placed on the table in front of the participant. The box on the left was green and labeled *Minor Challenge*. The box in the middle was yellow and labeled *Moderate Challenge*. The box on the right was red and labeled *Major Challenge*.

Below each box was an area where the participants could place cards. The decision makers would read a card that contained a barrier statement and, after reflecting on actual experiences at emergency scenes, would determine if that barrier had presented a minor challenge for them. If it had, the card would be placed in the green area. Likewise, cards denoting moderate challenges were placed in the yellow area, and cards indicating major challenges for decision makers were placed in the red area. To prevent participants from placing all the cards into one category, a rule was created that forced an equal distribution of the cards among

Fig. 5–4. A commander's situational awareness can be affected by the distraction created when another responder approaches. This is especially true if the information being shared is not mission critical.

the three choices. In other words, the participants were only allowed to put a certain number of cards into each category. The participants could change their minds and move the cards around as much as they wanted until they indicated they were done.

After the first sort there were 21 cards in the Major Challenge pile. All the cards in the Minor and Moderate pile were discarded from the exercise. Then the sorting exercise was repeated, using only the remaining 21 cards. On the second sorting exercise the participants were instructed to sort the 21 cards using the same Minor, Moderate, and Major Challenge criteria. Again, a similar rule was applied to force the distribution.

This resulted in seven cards in each category. The participants were then given the seven cards from the Major Challenge category and asked to sort them again into an order representing their most challenging to least challenging. This allowed me to see the participants' top three most challenging barriers. Finally, the participants were asked to rank their three cards (first, second, and third) from the previous sort. This process took the participants from a little more than 40 to just under 90 minutes to complete.

Interviewing expert decision makers

After the multiple card sort exercise was complete and the top three most challenging barriers were identified, the participants were interviewed to document their lived experiences with their most challenging barriers. The interviews addressed each of the top three barriers separately. Decision makers were asked to explain why they chose each barrier, describe how the barrier challenged them, and give examples of specific incidents where they had been impacted by each of their most challenging barriers. During the interview, participants were asked to offer advice that would possibly help other decision makers avoid having their situational awareness challenged by each barrier (figure 5–5).

Fig. 5–5. Expert fireground commanders were interviewed about their lived experiences with situational awareness barriers.

With the permission of each participant, the interviews were recorded, transcribed, and then coded. Coding entailed reading each line of the transcript to find instances where the decision maker mentioned any other barriers from the list. If the person talked about another barrier, that barrier number was listed in the margin. The number of times a commander referenced each barrier helped me understand how challenging these barriers are to good decision making.

In the process of coding the interviews, I learned that the participants identified new situational awareness barriers that I had not previously discovered in the existing research. These new barriers were identified as decision makers talked about how their situational awareness was challenged by their top three. I can confidently say that in each instance the participant were unaware they were offering up new barriers to be researched.

In addition to the discovery of the new barriers it also became very evident how interconnected many of the barriers were, meaning decision makers could actually be impacted by multiple barriers simultaneously.

For example, some decision makers described situations where their situational awareness had been impacted because of distractions (a barrier to situational awareness). They noted at the same time that they were frustrated or angry (which are two additional barriers to situational awareness) because of the circumstances that gave rise to a particular distraction. One such example was an emergency vehicle approached the scene with sirens blaring for no apparent reason and momentarily drew the decision maker's attention away from the incident.

The correlation between the barriers (the cause and effect of one barrier to another) was not evaluated in my study, though it presents a wonderful opportunity for additional research. (Hint to the doctoral student who may be reading this book while contemplating a dissertation on this topic).

Because there were so many barriers to situational awareness, it was admittedly difficult to get a handle on all of the data. To help manage this, I parsed the barriers into groups. Barriers with similar themes were packaged together and put into categories. For example, the category of "communications" included verbal communications, non-verbal communications, issues with progress and update reports, issues with misunderstood words and phrases, issues with incomplete communications loops, missed radio communications, radio equipment problems, incompatible radio equipment, using multiple radio channels (including cell phones), too much radio traffic, and crews not willing or unable to communicate.

The 12 situational awareness barrier categories

- Staffing
- Communications
- Data and information management
- Workload management
- Mental model
- Physical and mental stress
- Commander location
- Shared sense making
- Goals and mission
- Command support
- Human factors
- Attention management

The following chapters focus on the 12 barrier categories and discusses the most challenging barriers in each category, sprinkled with an occasional helping of the neuroscience behind what makes some of the barriers so difficult for first responders. Along the way, I share some interesting observations made by my research participants about other, less-often implicated barriers to situational awareness.

After discussing the barriers and explaining how decision makers have been impacted, part III focuses on the lessons learned from this (and other) research. It also examines the best practices that can help decision makers prevent being impacted from challenging situational awareness barriers as well as how to overcome them.

As tempting is it may be to jump ahead to the lessons and best practices, it's beneficial to understand the barriers in some detail before you are introduced to the solutions. Otherwise, I may liken the experience to a person who puts air in a flat tire, drives the vehicle until the tire goes flat again, and repeats the process. Absent the understanding of *why* the tire is going flat, the solution may seem so simple (put air in the tire). In reality, the *root cause* is something deeper (e.g., a nail in the tire), and that problem needs to be fixed before the tire will be fixed (hold air).

PART 2

BARRIERS TO COMMANDER SITUATIONAL AWARENESS

CHAPTER 6

STAFFING

Research shows that staffing issues can significantly challenge a decision maker's situational awareness. I can make this bold statement for several reasons. From the multiple-card sort exercise, 43% of the cards selected by participants were for staffing-related issues. The occurrence of short staffing in public safety agencies and the subsequent impact understaffing had on first responder situational awareness was a unique finding. In other words, as I evaluated the existing research on the barriers that challenge situational awareness in other high-risk, high-consequence decision-making environments, I did not find one instance where understaffing was identified as a barrier to situational awareness. In retrospect, this makes sense. It would be inconceivable that a commercial airliner would fly with just one pilot or a surgical team would undertake an operation without an anesthesiologist. Yet public safety responders are routinely called upon to perform with less than adequate staffing.

In such cases, a pilot's or a surgeon's decision-making environment is similar to other high-stress and life-or-death situations. But there's one key difference for first responders as it relates to staffing. If a surgical team or flight crew were understaffed, they would not engage. In other words, a short-staffed flight would be cancelled, or an understaffed surgery would be delayed.

The expert commanders in my research ranked situational-awareness issues related to staffing as their most significant barrier. The staffing-related barriers include: understaffing, unpredictable staffing, response time delays (as it relates to staffing), and lack of experience of the first responders they supervised. Although these issues can impact strategic and tactical perspectives as well, the focus here is on the decision making and situational awareness implications only.

Decision makers revealed that while they are responding to a call and upon arrival at the scene, part of their size-up considerations relate to staffing levels (both quantity and quality), including how long they think it's going to take to get the right staffing to the scene and how to manage the incident until the right quantity and quality of staffing arrives (figure 6–1). The intense focus on staffing is not surprising. Emergency scene operations are often labor intensive. It's no surprise that concern for risk to personnel would demand significant attention resources.

Fig. 6–1. Tracking the number of personnel responding to an incident is not enough. Commanders must also assess the quality of crews to ensure the tasks assigned can be completed in a timely fashion—a component of Level 3 situational awareness.

It's a commonly taught best practice for the first-arriving decision maker to conduct a size-up. A thorough and accurate size-up helps the decision maker develop strong situational awareness. As mentioned in previous chapters, situational awareness is developed by capturing cues and clues about what's happening in the environment and translating them into something meaningful (in the context of space and time). The responder then makes reasonable predictions about future events.

The initial size-up should also include an evaluation of the available staffing. For example, the decision maker might give consideration to the number of personnel responding to the call, how long it's going to take those responders to arrive at the scene, and the training and experience levels of the responders. Each piece of data about staffing represents clues and cues used for sizing up the staffing situation.

The decision maker evaluates the staffing and pieces all this information together into *something* meaningful—and that *something* is discerning if there are adequate resources at the scene to safely do the work that needs to be done. Then, consistent with the highest level of situational awareness, the decision maker constructs predictions about future events based on what the available personnel can get done within a given time frame, with consideration for the speed and complexity of the incident.

At a structure fire there is a finite amount of time in which responders can safely get the job done via interior operations (figure 6–2). At some point, a building on fire will deteriorate to a condition that makes it no longer safe for firefighters to be inside, or the smoke and fire conditions will become so intense that the interior environment is no longer compatible with life (regardless of how much protective gear is worn).

Arriving decision makers conduct a size-up that includes an assessment of on-scene and—as best they can—responding personnel. This may be easier for career departments who have fixed staffing levels and decision makers who know which responders are on duty and assigned to which vehicles at the start of the shift. Under these conditions, the decision maker can process the clues and cues about staffing well in advance of an incident and make plans to avoid compromising safety if a company is short staffed or there are crewmembers that lack training or experience.

Fig. 6–2. There is a finite amount of time during which responders can perform interior operations. At some point the conditions will deteriorate, and interior operations will no longer be safe.

In volunteer or paid-on-call departments where members respond from home to staff apparatus, the challenges of incorporating a staffing assessment into the size-up process can be challenging. Even if the officer on the apparatus announces on the radio the number of responding crew members, the decision maker has no way of knowing the training and experience levels of the crew members without actually knowing each person who makes up the crew on that apparatus.

Understaffing

When challenged by understaffing, a decision maker can't develop Level 1 situational awareness about staffing when critical clues and cues are missing, such as the quantity and quality of the responding crews (see chapter 4 for more about the three levels of situational awareness). Following the proper sequence of situational awareness development, it stands to reason a decision maker who lacks critical clues and cues can't progress to Level 2, deriving meaning from the combined clues and cues. In other words, the decision maker is challenged to develop a quality action plan without a thorough understanding of available staffing. Following this through to the final developmental level of situational awareness, Level 3, the decision maker may then struggle to understand how effective the crews will be toward achieving strategic objectives or tactical priorities because of the flaws in his or her Level 1 situational awareness. This example clearly demonstrates that the development of situational awareness is progressive—and to be effective, there can be no shortcuts.

Now, as with all information gathered in the formation of an action plan, the decision maker can make certain assumptions about the quantity and quality of the staffing levels in advance of their arrival. However, assuming a crew is capable of doing more than they can realistically accomplish or that the crew will be good enough to accomplish a task faster than is realistic can result in a flawed Level 3 situational awareness (being able to make accurate predictions about future events).

Far more common in police scenarios than in fire scenarios, an officer may be working alone during the initial operations (figure 6–3). This can significantly affect situational awareness, because the officer may be narrowly focused on one part of the emergency scene and there may be no one else watching out for the officer's safety. This can put the officer in a position of great jeopardy.

Fig. 6–3. Police officers are often required to initiate tactical operations without adequate staffing. This can increase stress and negatively affect situational awareness.

Overconfidence

Overconfidence by the decision maker about a crew's ability may cause a decision maker to over-assign tasks to personnel, thus giving crews multiple assignments and expecting all of the tasks to be accomplished in unrealistic time frames. Under the stress and excitement of an emergency this can be an easy trap to fall into. If the decision maker knows all the things that need to be done, the tasks can get assigned to the personnel present without consideration to crew size or ability.

Premature Assignments

Another challenge caused by staffing issues can occur when a decision maker arrives on the scene, conducts a size-up, figures out what's going on, and makes an action plan prior to the arrival of the proper number of responders to perform the assignments. Commanders reported to me that they have arrived on the scene, conducted good size-ups, and made a solid action plan, but they could not execute their plan properly because the right staffing had not yet arrived. And, unfortunately, some of these decision makers admitted that when the right quantity and quality of staffing finally did arrive, the decision maker gave orders to execute a plan that was formed in the past (sometimes five or more minutes earlier). Thus, the decision maker failed to give consideration to how the rapidly changing conditions should have changed the plan. In other words, the decision maker executed a stale plan. The clues and cues had changed, what was happening at the moment the staffing arrived was different, and what was going to happen in the future was also different. Executing a plan based on old information can be a dangerous decision. Yet the decision makers admitted they failed to realize the potential consequences of their actions.

Clearly, the decision makers weren't doing this on purpose. They gathered the information they needed, made assumptions to fill in the blanks (i.e., the missing information), and then developed and confidently executed their action plans. A decision making best practice might be to arrive and formulate situational awareness that includes a prediction of how far the fire is going to progress and how much the incident may deteriorate in the time it's going to realistically take to assemble the right amount of staffing. Then, as the staffing arrives, the decision maker can reassess the conditions to see if the predictions about where the incident would progress to before the staffing arrived was accurate or not. This type of time-lapsed continual size-up has to be conducted with a conscious understanding of the passage of time. In other words, the decision maker must pay attention to how much time is elapsing (seconds or minutes) as the incident intensifies. This helps the decision maker develop more realistic projections of how quickly the future events are going to unfold.

If a decision maker arrives, makes a plan, and fails to revisit the plan as staffing arrives, there is a risk the assignments set the responders up for failure because the current conditions and deterioration of the incident has already progressed beyond what is safe. If the decision maker stays true to an outdated action plan, changes in conditions may be overlooked.

Not All Staffing Is Created Equal

Nuances about staffing can challenge a decision maker's situational awareness in other ways as well. When crews arrive on the scene of an emergency there is a reasonable expectation that some action will take place. Depending on the culture of the organization, the decision maker may be challenged by crewmembers who expect to take aggressive action even when the decision maker knows the conditions are not conducive to aggression. (*Note:* organizational culture is a separate barrier to decision-making situational awareness and is discussed in detail in chapter 9.)

Response-time delays

Decision makers can also be challenged when the response times for emergency crews are delayed. It's stressful for a decision maker to await the arrival of tactical personnel on the scene of an emergency. This delay may cause decision makers to become frustrated or angry because they are action-oriented leaders, and their action plan can't be implemented until sufficient help arrives. Frustration and anger are barriers to situational awareness. The decision maker's attention may be honed in on where the crews are and why they haven't arrived yet.

Additionally, when a decision maker arrives before adequate staffing, a size-up and action plan must still be developed. If response time delays are substantial, the decision maker may conduct another size-up and revise their action plan. After conducting several size-ups and making several action plans, the crews arrive. The multiple size-ups and multiple action plans may cause bewilderment in the mind of the decision maker. The decision maker may suffer the ill effects of information overload as some of the facts (clues and cues) start to get jumbled up in the decision maker's mind, leading to confusion about what information was part of the first size-up, versus the second or the third.

Quality of staffing

Another way staffing can challenge a decision maker's situational awareness comes from issues relating to the quality of staffing. For example, a decision maker responding to an emergency may formulate a mental model of what is happening at the scene based on the information that's provided by the dispatcher as well as the decision maker's existing knowledge from pre-incident planning or prior knowledge. The decision maker then arrives at the scene, conducts a comprehensive size-up, and knows what needs to be done.

One criterion for a decision maker's size-up should be an assessment of the quality of the personnel (figure 6–4). As the crews arrive, the decision maker assesses the training and experience levels of each crewmember. It's not enough that the decision maker knows the crew has three, four, or five members because the competency of a crew is based as much, if not more, on the *quality* of the crew as *quantity* of the crew.

Decision makers who assess crew quantity and quality prior to making a tactical assignment can improve situational awareness. Occasionally, through no fault of their own, decision makers can be dealt a proverbial bad hand of cards. If there's one thing you learn from gambling, it's you

Fig. 6–4. The age and experience levels of firefighters are clues and cues that help a commander form situational awareness about what they can accomplish and how long it will take them to accomplish it. It is unrealistic to hold the same expectations for an inexperienced crew as are held for an experienced crew.

don't bet large sums of money (take big risks) when you have bad cards. Sure, you can try to bluff your opponent. But in cards, your opponent is another human who is also trying to assess your motives and outthink you. Fire, on the other hand, doesn't care if you have a strong hand or a weak hand. It's going to burn indiscriminately. Your first defense is a strong offense. If you don't have a strong offense, contemplate the benefit (and the consequence) of engaging in high-risk operations.

Some decision makers, despite knowing they have a weak crew, still engage in high-risk operations. Such a decision can have a significant impact on the decision maker's situational awareness. As risk taking increases and crew safety is compromised, the attention of the decision maker can become narrowly focused on the safety of the crews at risk.

In the process of this focus narrowing, the decision maker can become oblivious to other things happening in the environment. Critical clues and cues can be missed because of the decision maker's hyper vigilance to the crews who are operating in an environment they (arguably) perhaps should not be in. The decision maker realizes the safety of the crews has become a gamble, which causes increased stress. The decision maker is no longer in control of the outcome. Compare this to any situation in which people gamble on an outcome whose variables they lack control over—such as in any casino game. This is a dangerous position to be in because the crews are at risk right from the start of the operation, and they need the decision maker looking out for their safety. The decision maker needs to maintain a big-picture view of the incident, being ever aware of the changing conditions. Stress from taking excessive risk narrows attention and affects situational awareness.

Task focus commanders

On occasion decision makers may find themselves *going operational* (being hands-on) when it's determined the crew lacks sufficient training or experience (figure 6–5). In this situation, it's a judgment call that may, in reality, improve crew safety. However, going operational can also increase the risk to personnel. When a decision maker goes operational there's no longer anyone commanding the incident. This can cause the decision maker to have a narrow focus and can most certainly cause big-picture clues and cues to be missed. Going hands-on can also cause the decision maker's focus to become fixated on the safety of the inexperienced crew.

So, which is better: Sending an inexperienced crew into a hazardous environment without proper supervision, or ensuring someone maintains a big-picture view of changing conditions? The question is easy to ask but the answer is difficult to put into an absolute recommendation. Suffice it to say that either decision has an element of risk associated with it. There is a third option: Don't engage an inexperienced crew in a high-risk operation.

Hopefully some decision makers reading this section are thinking to themselves, "*I would never engage in high-risk operations with undertrained or inexperienced personnel*." To that I say, congratulations! The likelihood of

Fig. 6–5. Occasionally, staffing levels may create a need for a commander to become operational. While this may be done in the best interest of crew safety, efforts should be made to ensure an incident has a dedicated, hands-off commander as soon as possible.

incident-scene casualties is vastly reduced when a decision is made *not* to engage in high-risk operations because of the quality or quantity of the crew.

So why would decision makers engage in high-risk, high-consequence operations when they don't have enough personnel on the scene or when the personnel they do have on the scene lack the training or experience to do the job safely? We explore the answers to these questions in the coming chapters.

An observation about staffing

I want to pause here for a moment to share something else I learned while conducting my research. Staffing, as a barrier to situational awareness, may be a finding unique to public safety responders. This may be a somewhat unqualified statement because I admit that I have not evaluated every study ever conducted on situational awareness in high-risk, high-consequence work environments. What I can say with authority is that I have read hundreds of them. Those studies formed the foundation of my research on situational awareness in public safety domains. I evaluated the barriers that impact fighter and commercial airline pilots, surgical and emergency room teams, air traffic controllers, submarine commanders, and many others who work in stressful environments similar to public safety.

In evaluating all of the studies, issues relating to staffing—understaffing, delays in the arrival of adequate staffing, undertrained and inexperienced personnel—were never mentioned. Lots of other potential barriers were, such as distractions, interruptions, poor communications, and information overload. But staffing issues were not among them. This finding intrigued me. Why had this never before been identified and discussed? I reviewed additional literature. I still found no other studies had been conducted.

Thus I conclude the impact of staffing on the situational awareness of a decision maker may be a unique finding to public safety domains. I realize there are other professions and circumstances where staffing issues affect situational awareness. I also think the impact that staffing issues has on public safety decision makers may be particularly challenging because of the speed with which decision makers must formulate and maintain their situational awareness under stress. For the moment, this remains a unique finding from my research. I challenge other researchers who may read this book to pick up this issue and advance our understanding of the staffing impact on situational awareness by applying this finding to other high-stress, high-consequence domains.

CHAPTER 7

COMMUNICATIONS

Recall from our discussion in chapter 5 the list of 116 situational awareness barriers were parsed into 12 categories. The second category of barriers deals with issues related to communications. If staffing (discussed in the previous chapter) is an issue unique to public safety, then communications is an issue universal to every domain where high-risk, high-consequence, high-stress decisions are made. This shouldn't be too surprising when you think about how challenging it can be to communicate effectively with people in low-risk, low-stress, low-consequence environments. Under the best of conditions there can be miscommunications. When stress, uncertainty, changing conditions, and risk are present, communications can become extremely challenging.

You Don't Know What You're Missing

Recall from chapter 4 that the first level in the development of situational awareness is capturing the clues and cues in your environment. This includes capturing verbal clues and cues from dispatchers and responders. To miss radio communications is to miss clues and cues. How do you know if the information you miss is important? The answer to that question is usually made apparent during the post-incident evaluation. Missing radio communications can lead to additional communications issues, too. This is especially problematic when the sender of the communication thinks the message was received when, in reality, it wasn't. This may happen in haste when a sender delivers a message and never waits for acknowledgment from the receiver. The sender assumes receipt of the message; this leads to a disconnected understanding between personnel who are supposed to be on the same page, working toward a common goal.

A communications disconnect between the decision maker and the crew is the equivalent of an orchestra where the wind section is playing Beethoven, while the strings are playing Garth Brooks and percussion is playing Snoop Dog. It's not going to sound pretty. On the emergency scene, however, the consequences can be catastrophic. As noted, the importance of missed radio communications may not be readily apparent (because the decision maker does not realize the message was missed). At some point, though, that missed message may become very important to the safety of first responders.

Imagine you go to the movies with friends. After you enter the theater and find your seats, your companions decide they'd like some popcorn. You offer to go get it. The previews of upcoming movies just started so you know you have at least 10 minutes to get the snacks and return.

The lines are longer than you expected, and while you are out of the theater, the movie starts. The first scene in the movie is a lady walking to her car at night. She fumbles through her purse for her car keys. She stumbles and the keys fall out of her purse on to the ground. She stops and bends over to pick up her keys. As she does, the contents of her purse spills out on the ground. She quickly shovels the contents back into her purse. Keys in hand, she unlocks the car door, gets in, and drives away. As she is pulling away, you return from getting the popcorn and take your seat. Disappointed that the movie started while you were out of the theater, you ask your companions what you missed. They tell you the movie just started and you didn't miss anything. You're relieved.

About 40 minutes into the movie, something significant happens and you're confused. You don't understand what it means. This is because, unbeknownst to you (and your companions), when that lady was fumbling through her purse at the start of the movie something besides the car keys fell out of her purse on to the ground. The clue was so subtle that your companions barely noticed it and surely didn't think it was important enough to mention it to you. Now, suddenly and unexpectedly, that's a critical clue. It seemed so innocuous, so inconsequential at the time. Now you're confused and maybe a little frustrated, because you're clueless as to what's going on. This is what happens when you miss critical radio traffic or, for that matter, any critical clue or cue. The problem is you don't know what a critical clue or cue is at the time. That lesson usually comes after the fact . . . hopefully not as part of a post-incident evaluation of a casualty incident.

As challenging as missed radio communications are, the lack of update or progress reports can be equally as daunting for decision makers. There are several reasons this happens. For example, crews simply may not tell the decision maker what is going on. This may occur because the responders may erroneously assume the decision maker can see what the conditions are like. Sometimes the conditions are visible from the command post, but that's not something that should be assumed or left to chance. Speaking from personal experience, there have been numerous occasions where I had crews come out of a building on fire and tell me how hot it was inside and the evidence was their scorched gear or damaged helmets (figure 7–1). From the outside, I had no idea that the conditions were that bad on the inside. Had I known, I would have ordered them to come out. They were taking a beating and didn't tell me.

So why wouldn't a crew provide a progress report? There are several possible explanations. Maybe they don't know how to talk on the radio, and they don't want to sound stupid by stuttering and stammering through a progress report. Maybe they're concerned they'll say the wrong thing. Maybe they didn't have a working radio, or their radio was tucked in a pocket and they couldn't reach it. Maybe their radio was inadvertently on the wrong channel. Maybe they were too busy trying to perform operational tasks to talk on the radio, or it's their department's culture (or policy) for operational personnel to stay off the radio. Whatever reason, updates and progress reports were nonexistent. Absent those clues and cues, the crews were at risk of being on a different page than the decision maker. In other words, the crews and command didn't have a shared (common) situational awareness.

Fig. 7–1. Sometimes the only indicator of how hot it was on crews operating inside a structure fire is seeing their scorched gear when they exit the building.

Another communications-related barrier is an incomplete communications loop. This can happen when a decision maker gives an order to an operational crew and, instead of repeating back a summary version of the order and confirming their understanding, the crew simply responds with "10-4" or "copy" or "message received." The decision maker then assumes the order was received and understood exactly as it was intended, but in many cases it wasn't. There was a miscommunication, but no one knew it. This can be made worse if the sender misspeaks, uses words or phrases that are unfamiliar to the receiver, or uses words or phrases that can have more than one meaning. Think about how many times during normal, low-stress, low-risk conversations you have had this happen and you can begin to see it's not difficult to have this type of miscommunication. Adding stress compounds the potential.

There are several ways to improve communications. One way is for the decision maker to never miss the communications from their most at-risk personnel, which usually includes the personnel operating in the high hazard area (or hot zone)—the environments where if something goes wrong the personnel may only get one chance to transmit a critical distress message.

Another way to improve communications is to ensure the use of a complete communications loop (figure 7–2).

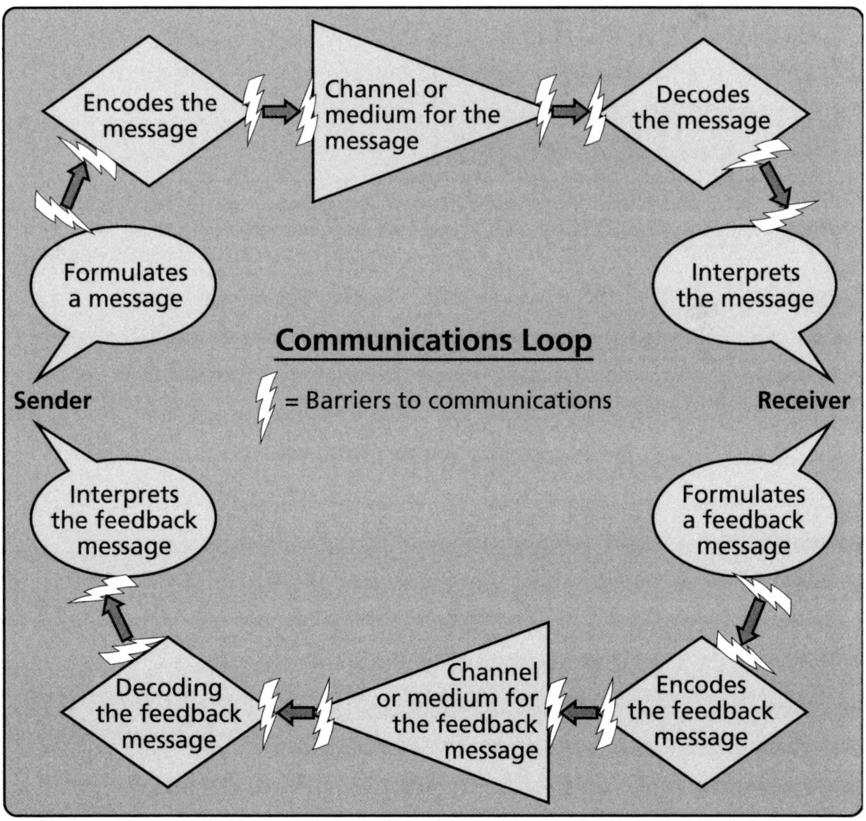

Fig. 7–2. The communications loop is fraught with opportunities for barriers to impede the quality of the message. There are many barriers that can negatively affect understanding.

The concept of the communications loop is taught routinely in business communications courses. The sender transmits a message. The receiver captures the message and processes it into meaning. The receiver then formulates a confirmatory message and transmits it back to the sender. If the message is correct, the sender confirms it. If the message is not correct, the sender can make the correction. After the receiver confirms the common understanding of the intended message, the communications loop is complete.

Here's an example:

> COMMAND: "Engine 1 from command."
>
> ENGINE 1: "Command, this is Engine 1."
>
> COMMAND: "Engine 1, I want you to advance an inch-and-three-quarter line through the back door to the second floor and conduct a search and attack."
>
> ENGINE 1: "Command from Engine 1, copy an inch-and-three-quarter line, through the front door to the second floor for attack and search." (Note that Engine 1's repeat of the message was incorrect, causing command to correct the error.)
>
> COMMAND: "Engine 1 from command, negative. I want the line through the back door."
>
> ENGINE 1: "Command from Engine 1, copy line through the back door."

Communications can also be improved when the sender and receiver use commonly understood terminology, words, and phrases. This sounds simple enough but is often overlooked or taken for granted. Under stress, responders may struggle with their words. If communicators are winded or in a hurry, they may also use shortcut phrases. If you know this is going to happen, take pre-emptive steps to address it. Develop words and phrases for the most common tasks, assignments, and updates.

After you develop common words and phrases, practice them over and over again until they become habit. Practice them in training, on small incidents, and during idle time. There is no such thing as over-practicing, especially when it comes to performing under stress. There will be times when you get only one opportunity to get it right, and lives depend on clear communications. Under such conditions you need to establish good habits. The messages you need to convey and how the messages from others are interpreted are critical to the safety and success of emergency operations.

Another communications-related barrier that can challenge decision makers is too much radio traffic. Keep in mind that every radio transmission has to be processed into meaning. Those messages consume precious working memory—memory decision makers need to figure out what is going on and what is going to happen next. Critical messages facilitate understanding. Noncritical messages (read: noise) hinder the decision maker's ability to think and develop clear mental models about what is going to happen next (i.e., Level 3 situational awareness is flawed).

Decision makers I've talked with have often expressed frustration over too much radio traffic. They say things like "Everyone who has a radio thinks they have to talk on it." Or, "Our department issued everyone a radio, and that was a big mistake because now you can't get a word in for all the chatter."

The brain has no way to determine a critical message from a garbage message until it processes it. That processing can require some heavy mental lifting and consume precious cognitive resources. There's no holding queue for audible messages. You either hear it and process it, or you miss it (figure 7–3).

There are some other communications-related barriers that warrant discussion. Radio equipment problems can create barriers to communications, especially when they go undetected. If a radio is not functioning, or the radio channel control dial gets bumped and the channel is switched, the crew in the high-risk environment may not know it happened. There could be critical messages being communicated from command to the crews on the incident scene and a crew with radio problems may not know it. Thus, they will not be able to capture clues and cues, and their Level 1 situational awareness will be flawed.

Fig. 7–3. When a large amount of traffic is transmitted via radio, it can have an impact on situational awareness.

Another communications barrier can arise when multiple departments responding to a call have incompatible radios. The information being shared over the radio, whether the source is from dispatchers sharing caller information, from first arriving companies providing size-up information, or from decision makers giving initial assignments, these are all critical clues and cues. If you don't have compatible radio equipment, responding personnel are not going to be on the same page when they arrive. Thus, there is no shared understanding, and the incident-wide situational awareness is flawed.

Stated more directly, crews will not have a commonly understood sense of what is going on and what needs to be done. If mutual aid arrives on a scene after missing critical information shared over incompatible radios, there's a chance they are going to be confused and frustrated (both of which are barriers to maintaining situational awareness).

But the problem doesn't end there. If you're going to try to rectify the problem in real time at an incident scene, someone has to be responsible for conducting a briefing. Someone has to bring the mutual aid responders up to speed. That *someone* is often the decision maker. This means that in addition to performing all the duties of being the incident commander, the decision maker has to surrender some precious mental resources to pause, think about what the incoming mutual aid personnel needs to know, form the message in a fashion so that when it is articulated it makes sense, and then actually talk to the mutual aid responders. Depending on how many mutual aid responders are coming that do not have compatible radio equipment, this can become a formidable barrier to the decision maker's situational awareness. This is likely to play out during a time of rapidly changing conditions—a time when the decision maker may have no excess capacity to commit to this task. So something has to give—either the mutual aid responders don't get all the information they need, or the decision maker fails to maintain proper situational awareness of the incident scene while repeatedly providing these incident briefings.

Decision makers can also be impacted when incident operations are assigned to multiple radio channels. As much as you may think you can multitask duties at an incident scene, the human brain cannot multitask the act of listening and you will struggle to hear and comprehend multiple conversations at the same time. And, when multiple people are talking, your ability to comprehend any conversation is diminished unless you intently concentrate on what one person is saying.

> *Just because someone is talking doesn't mean you are always listening.*

Emergency services dispatchers have told me they can acquire the ability, through practice, to effectively listen to and comprehend multiple conversations simultaneously (figure 7–4). Short of testing them I'm not in a position to dispute their claims, but it does seem plausible that with practice, any shortcoming can be overcome. I hypothesize they're not actually multitasking the act of listening. Rather, in a quiet, consistent environment (like a dispatch center), while wearing headsets and concentrating on the act of listening, a dispatcher can learn to pick up on keywords and voice inflections that tip them off to listen to one conversation over another.

Fig. 7–4. Dispatchers wearing radio headsets while sitting in a consistent environment may be better suited to monitor multiple radio channels. The relatively quiet environment may improve concentration and the ability of a dispatcher to switch attention to different channels more quickly.

While conducting some research on communications challenges, I got to observe air traffic controllers at a busy international airport. The controllers wore radio earpieces that fit into one ear and had a microphone attached to them. The other ear was open to listen to the conversations of other air traffic controllers in the room. I observed the controllers listening to the conversations through their radio earpieces while seemingly carrying on conversations with other controllers in the room. I asked a veteran controller (a gentleman who had more than 30 years of air traffic management experience) if he could effectively listen to and comprehend multiple conversations at the same time. He told me that he could not. He did say there are certain key words or phrases that capture his attention and cause him to shift his attention to the other conversation. He also admitted, however, it doesn't always happen that he hears those key words or phrases when they are said the first time.

I think it would be a rare skill if a person could simultaneously listen to and comprehend multiple conversations at the same time. As a general rule, I would recommend you assume you cannot and take steps to ensure you are not in a position to have to do so. Otherwise, you may be fooling yourself—or worse, missing critical clues and cues and be setting yourself up for a near-miss or catastrophic event.

Another communications-related barrier that you should be aware of is missing radio communications from operating on multiple radio channels. There's nothing wrong with operating an incident on multiple radio channels. In fact, there are some logical reasons for doing so, including the management of work groups that may require extensive radio traffic. However, if a work group is under the assumption that supervisory personnel (e.g., the sector/division officer, operations officer, or the incident commander) are monitoring their traffic when they are not, there can be problems. Or, if supervisory personnel assume the work group is monitoring the supervisor's communications when they are not, this can also cause problems. When two people are communicating, whether in person, by telephone, or over the radio, it's fair to assume both parties are listening and understanding what's being said. If either of the parties is not listening, then the sender and the receiver are not on the same page (i.e., they will not have a shared understanding of what is going on). This problem can be compounded when work groups and supervisors incorrectly assume that everyone is listening to and understanding what is being communicated, when in reality some may not be and there may not be a common understanding of who is on the same page and who isn't. It's not difficult to see the recipe for disaster that's brewing.

Communications can also be hindered when a responder is unable to communicate by radio. There are a variety of reasons why this can occur. It can be as simple as the radio is located in a pocket that's not easily accessible. Maybe the user has to remove the radio from their pocket to talk. If so, maybe the pocket is blocked by an SCBA strap or a shoulder strap being used as a sling for a flashlight or thermal imaging camera.

In some organizations portable radios are not available for all personnel operating in hazardous environments. This may be a result of budget issues, or it may be an administrative decision. Regardless of the cause, a responder in a hazardous environment without the means to communicate (both as a sender and a receiver) is in a dangerous predicament, and situational awareness will be impacted.

Some responders have reported they refrain from talking on their radios because it violates a cultural norm in their department. "Personnel should be working, not talking on the radio" is the mind set of their members. The challenges of an organization's culture are discussed in greater detail in an upcoming chapter.

Next, we look at how decision making and situational awareness are impacted by issues related to physical and mental stress.

CHAPTER 8

PHYSICAL AND MENTAL STRESS

Emergency incident scenes are inherently stressful, both physically and mentally. Rarely do responders feel underwhelmed (bored) at an emergency scene, though it has been reported that boredom, manifested in the form of complacency, can occur when responders let their guards down. This will be discussed later.

> *Any responder not feeling some stress at a scene is either operating at a very minor incident, or they are not grounded in the reality of all that can go wrong.*

At the scene of an emergency, information is changing rapidly and there are lots of unknowns. Personnel are at risk of being injured or killed, and the incident is unfolding rapidly. Responders are far more likely to feel overwhelmed from all that is happening around them. Under these conditions, decision makers may experience a variety of emotion-triggered responses including frustration, anxiety, confusion, anger, and excitement. This can impact a responder's ability to process information. The greater the stress, the harder it becomes to process information—especially complex and novel information (figure 8–1).

Fig. 8–1. The more dynamically changing an environment is and the more risk responders must assume, the greater the stress will be on commanders.

Although your brain has some amazing abilities, it also has limitations. Some of those limitations relate to the volume of complex and novel information it can receive and process. You may also recall from discussions in an earlier chapter that when under stress, your brain may revert to using intuitive decision-making processes versus rational processes. This can add to the stress.

Let's Be Rational

You live in a rational world where you're taught to be a rational thinker. When you make a decision you often support it with facts, data, statistics, charts, graphs, and so on that provide the proof (evidence, if you will) that you're making a good decision. Unfortunately, emergency scenes are not very forthcoming with facts, data, statistics, charts, and graphs (perhaps we should count this among our blessings). Whether you consider it a blessing or a curse, the lack of information and rapidly changing information can cause stress. If you're an analytical person who needs complete reassurance of your facts prior to making a decision, you may feel some stress. The sheer volume of information you have to process, the speed in which it is coming at you, how fast the facts are degrading, and conflicts in factual information may all contribute to a responder's stress level. As you get more overloaded, overwhelmed, and frustrated, your stress level may increase. This may challenge your ability to think things through and make good decisions. This creates a special concern as it relates to your situational awareness.

Recall Level 3 situational awareness provides you with the ability to project future events. Under stress, your brain may be focused on more immediate (i.e., present moment) priorities. Recall also that when under stress, your brain's attention is driven by hereditary instincts—with the most powerful attention grabbers relating to survival. When stress levels elevate, your brain gives preference to processing the information essential for survival in the current moment.

This isn't a bad thing. In fact, it can save your life. However, as often is the case in first responder casualty incident reports, the outcome seems so predictable based on how the events are unfolding. In other words, as you read a case study, near-miss report, or the line-of-duty death report, you may find yourself able to readily predict the future—to see the bad event coming. The reason you can do this is two fold. First, you already know the outcome; thus, you have *hindsight bias* because you know to look for the indicators that something bad is going to happen. I discuss this in more detail later on in the book when I discuss best practices for learning from the mistakes of others. Second—and this is the important point as it relates to the topic of this chapter—when you read a fatality report, it's not likely that you're doing so in a high-stress environment. As you read the case study document, you're probably sitting in an office or a training room or in the comfort of your home. Because you're not under stress, your body is not on high alert for danger. Your senses are not scanning the environment in search of clues and cues of things that can kill you. Thus, it's much easier for you to capture and comprehend the indicators of the future events as you read the report.

This is one of the traps first responders can fall into. In a low-stress environment you may find yourself reading a case study where the incident is degrading quickly. The low-stress environment makes it much easier for you to see clues and cues that indicate a bad outcome is on the horizon. We can, however, fail to understand that the reason the deteriorating situation seems so obvious is *not* because of the overt presence of the clues and cues (perhaps seemingly screaming out that things are going bad). Rather, it is because you're reading a report in a calm environment, devoid of the real-world stresses and pressures that were present at the actual incident scene. This can give you a false confidence that if *you* were at the scene, you would not have overlooked such obvious indicators. Try to avoid being lulled into the false belief. Even the most experienced responders can miss critical clues and cues under stress (figure 8–2).

Fig. 8–2. Reading a first responder casualty report in a calm environment gives the reader a distinct environmental advantage—no stress. This can help the reader see critical clues and cues that were not as obvious at the high-stress emergency scene. When this happens, the reader may pass judgment on the decisions made at the emergency scene, viewing them as flawed or lacking good judgment.

What is obvious to you in a case study may not be as obvious to you if you were operating at the actual emergency scene. Many of the responders operating on the scenes of these emergencies where bad things happened were smart responders with lots of training and decades of experience. If stress can cause them to miss important cues and clues that result in a catastrophe, then it can happen to any of us.

CHAPTER 9

WORKLOAD MANAGEMENT

In the aftermath of catastrophic events, emergency scene commanders often confide that they were affected by an excessive workload or poor management of their workload. Emergency scenes are complex environments where dozens (and perhaps hundreds) of pieces of information must be processed. Complicating this is the speed at which the information changes and the need to develop and execute a plan of action against a situation that is continually morphing into something different.

Imagine for a moment participating in an experiment where you are provided with some basic information about a problem you need to solve. Then you are given a single sheet of paper containing 50 facts, some of which are vitally important to your problem and others that are not so important (but it may not, at this point, be entirely obvious to you which facts are important and which are not until the experiment plays out). When you're done reading the fact sheet, you have to develop and implement a plan of action. When you're halfway through reading the fact sheet, the moderator hands you a new two-page fact sheet and warns you that some of the information has changed since the first fact sheet was printed. However, the differences are not bolded or highlighted; you have to compare the sheets to figure out which facts have changed and which have not. Then, while you're in the middle of figuring all that out, the moderator hands you a four-page fact sheet and once again tells you some of the facts have changed again.

So let's recap. You now have before you three sets of facts on one-, two-, and four-page documents. Some of the facts on the first page stay consistent and true on all three documents. Some of the facts on the first document change on the second document, but not on the third. Some of the facts on the second document remain the same on the third document, but some change, too. Each new document may contain new information that the previous one did not. Got it? Good. Each time you are handed a document you're given five minutes to figure it all out. Can you see how easy it is to become overwhelmed with the sheer volume of information and the speed at which you have to process and make sense of the changing information? Although this experiment was a hypothetical event, managing an emergency scene is very similar in nature.

Easily the most sophisticated information storage device known to us, your brain has an amazing capacity for storage and retrieval. However, your brain has a limited capacity for how fast it can input and process new information. Similar to a personal computer, when you try to input a large amount of data, it utilizes all the capacity of your short-term memory. If this happens on your home computer, you get visual prompt on your screen. For example, if you use a Microsoft product, you get the little hourglass with the sands of time passing through it. If you use an Apple product, you get the psychedelic spinning circle. These prompts tell you the computer is busy. At the point your computer (or your short-term memory) reaches capacity, you can cancel the data load, because your computer is not going to accept any more information.

The same sort of event can happen when your brain is using all of its capacity to process information about the emergency scene. Your "hourglass" icon pops up, and your brain stops processing new information because it has reached its maximum capacity. When this happens, new information may not get in, literally. Your eyes may not see visual cues and clues. Your ears may not hear audible clues and cues, including important radio messages. In the most literal sense, your eyes will see (so long as your eyelids are not shut) and your ears will hear (so long as you don't have ear plugs in), but your brain may not process the visual and audible cues and clues because it's too busy trying to process other information.

This might explain, in part, how the clues and cues leading up to a catastrophic event can unfold right in front of responders. While they are looking right at the scene or listening to it, they may not be able to see or hear what's happening around them. This neurological shortcoming might help to explain documented cases where at-risk crews transmitted multiple radio distress messages and the incident commander reported having never heard the messages. The amount of information processed, the speed at which that information changes, and the complexity of the emergency scene can all contribute to the possibility of overload.

I'm Good at Multitasking

Some responders believe they can efficiently and effectively perform multiple roles and multitask on emergency scenes. This belief scares me because responders can get a false sense of confidence about their ability to multitask. There is a growing body of evidence, supported by functional magnetic resonance imaging technology (fMRI), that proves it's biologically impossible to pay attention to more than one thing at a time.

Some of you will find this difficult to believe; others will dismiss it without a second thought because you believe you are good at multitasking. Stick with me and let me explain it in a little more detail, and I think you'll understand what science is proving. Everyone multitasks. If you are walking and talking, you are multitasking. If you are driving your car and listening to the radio, you are multitasking.

Think of your brain as having a two-part attention management system—an autopilot and pilot system, if you will. The first part, the autopilot, is your subconscious brain. At the subconscious level your brain stores learned behaviors and allows you to perform tasks without giving them much thought (like walking, eating, and for the most part, driving). Think of these as tasks that happen automatically—on autopilot, if you will. Your brain still thinks about how to do them, but because you have done them so many times before, your subconscious brain is able to control the behavior from behind the scenes. This is similar to a personal computer that runs programs (like virus scanners) in the background while you play video games or chat with a friend online on the screen.

The second part of your brain that guides your actions is the pilot—your conscious brain. Physical activities that are new to you or that are not memorized are controlled by your conscious brain. In other words, to do these tasks, you have to actually think about (and concentrate on) what you're doing.

The physical activities you do all the time (like walk and drive a vehicle) can be done mostly without conscious thought; you do these things on autopilot. The physical tasks you do less frequently (perhaps pulling a hoseline or setting up a ladder) require conscious thought, because your brain doesn't have the physical movements memorized. It's within the context of your conscious attention where the notion of multitasking can go awry.

Because many of the actions you do are driven by subconscious programming, and you're so good at doing multiple activities that are programmed, it's easy to believe you can multitask even when performing activities that require conscious awareness. This is where the problems begin.

Multitasking at the conscious level is impossible to do; you can only focus on one thing at a time. However, you can quickly shift your focus from one thing to another, to another, and back to the first in very rapid succession—so fast, in fact, it may seem that you are multitasking. What you are doing is single tasking in a very rapid succession back and forth, focusing on one visual cue, then the other, then back to the first.

Who cares? What's the big deal so long as the job gets done? Right? Here's the "big deal." Every time your conscious brain shifts from one attention task to another and then back to the first task, there is a chance that you will either forget where you left off or struggle to locate the point of reference where you left off. This can make you vulnerable to picking up at a different spot and, thus, forgetting to do something important.

There is a very quick exercise you can do that illustrates the multitasking limitations point well. Hold this book at arm's length with one hand, and read the words on the page (assuming your eyesight allows you the luxury to do so). Now take the index finger of your other hand and hold it about 6 inches in front of you in such a fashion that it is between your face and the book. Now, focus on the book so that you can read the words and notice that your finger will become blurred (or you may even have double vision—seeing two of your finger). If you adjust your eyes to bring your finger into clear focus, the words of the book become blurred. Now try to multitask and focus on both the book and your finger at the same time (figure 9–1). What happened? Were you able to clearly focus on both? Finally, focus on a particular sentence in the book, then focus on your finger, then back to that sentence in the book, then back to your finger, and back to that sentence in the book. Notice how easily your eyes were able to adjust from one task to the other, yet it was impossible to perform both simultaneously.

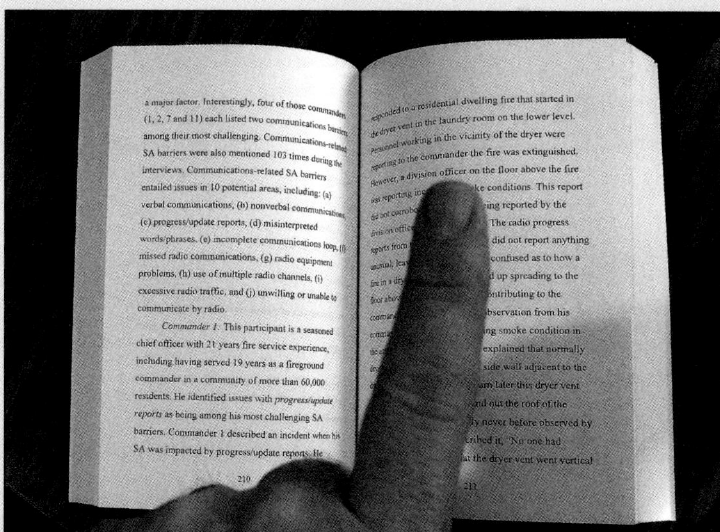

Fig. 9–1. If you hold your finger six inches away from a book and focus on your fingertip, the words will be blurry. If you focus on the words, your fingertip will be blurry. This is because the eyes cannot multitask when it comes to focusing. The same is true for attention. You cannot pay attention to two things at once.

When you went through the finger/book focus exercise, did you have any difficulty coming back to the very sentence—or the very word—where you left off? Maybe you did, maybe you didn't. Remember, you're not under any stress during this exercise; the conditions are not dynamically changing around you. When you're on an emergency scene, the conditions are very different.

Conscious Awareness

For many first responders the activities performed on the emergency scene must be done using conscious memory. There are two reasons for this. First is the lack of continual and repetitive practice. Repetition of tasks can lead to memorization. When this happens, the knowledge of what to do and how to do it transitions from conscious memory to subconscious memory. Using the previous computer analogy, the information goes from the desktop (short-term memory) to the hard drive (long-term memory). Tasks residing in long-term memory can be performed subconsciously. Second is the notion that every emergency scene is uniquely different, and this presents the responder with a novel set of circumstances that cannot be managed using subconsciously stored information.

> *Allow me to state it as simply as I can. The more tasks you try to perform simultaneously under stress, the more likely you are to make mistakes. The more information you try to process in your conscious memory, the more likely you are to become overloaded and overwhelmed.*

What you don't know, in advance, is what information you're going to forget or what task you're going to inadvertently fail to do and the consequence of that oversight. Those are the lessons that embarrassingly come out in after-action reviews.

Performing multiple roles

Think about an emergency scene for a moment, and make a list of all the roles that need to be filled. Now make a list of all the duties that go along with each role. That makes for a very long list of duties and responsibilities. If you perform multiple roles, then you own the duties and responsibilities of each. However, a dynamically changing, high-stress emergency scene will rarely accommodate your ability to completely focus on one role at a time. Instead, it's often the case that the responder who performs multiple roles will have to alternate between the duties of one role and the duties of another (assuming the person is performing only two roles).

For example, if you're the officer on the first-arriving company at a residential dwelling fire, what role(s) do you assume?

A. Incident commander

B. Division leader

C. Safety officer

D. Operations officer

E. Company/group leader

F. Logistics officer

G. Liaison officer

H. All of the above

I. More than one, but which ones and how many vary depending on staffing and the complexity of the incident

If your answer is H or I, then you need to understand you are at risk of becoming overloaded and overwhelmed because of the limitations of your brain to process information at the conscious level.

When you multitask at the conscious level (which, you may recall, is actually single tasking in multiple succession), you are setting something aside, if only momentarily, to do something else. When this happens two results can happen. Remember the finger and book experiment earlier? First, you can forget where you left off. Second, you can come back to the task you were performing and think you already did a step when, in fact, you had only thought about doing it and you actually never did it. Regardless of the means, the end result is the same—the potential of failing to do something important.

If the oversight results in an injury or fatality, the lesson is painfully immediate. If you're lucky enough that the oversight is not mission-critical, then the shortcoming may not be revealed until the post-incident evaluation (if at all). Responders have confided in me about several examples of this shortcoming. In one case, a responder told me, "I asked my supervisor to do something for me, and he acknowledged the request and informed me it would be taken care of, and it didn't happen." When the responder called out the supervisor for the error, the supervisor's response was, "I remember you asking me to do that, and I remember starting to do it for you. Then I was drawn in the direction of doing something else, and I never got back to doing what you asked me to do." Oftentimes supervisors cannot even recall what drew their attention away in the first place. The reason is secondary to the potential consequence of such an oversight.

Physical tasks are mentally demanding

Whenever you perform hands-on tasks, (especially of the kind that are routine and don't reside in your subconscious), the demands on your brain are significant. This is because your brain has to think about both the steps of the task and to coordinate all your eye, ear, equilibrium, and muscle movements in unison (lest you stumble around like a buffoon).

If you're in the role of the commander and are performing hands-on tasks, something is likely to be overlooked. Most often, that something is going to be a higher-level thinking task while your brain is concentrating on coordinating motor movements. It's part of your hereditary instincts—first priority is given to performing physical activities that keep you from getting hurt or killed.

The Importance of Being a "Hands-Off" Commander

It is important for responders to understand that the instant a commander takes on performing a physical task, whatever that task is, his or her ability to process and understand what is happening at the overall incident scene can be substantially affected. This happens two ways: your need to use precious brain capacity to perform the physical task; and your inability to focus attention on two different things at the same time (figure 9–2).

Fig. 9–2. Performing physical tasks on an emergency scene, such as coupling a hoseline, requires a lot of mental energy. If a responder engages in a hands-on task, his or her ability to focus on the big picture and process critical clues and cues can be impaired.

You simply cannot be hands-on and focused on the big picture simultaneously.

If you try, you're going to miss something. It's challenging enough for commanders to see, hear, process, and comprehend things that are happening in high-stress environments, even when their complete and total attention is given to the task of paying attention. Complicating the command role with hands-on duties is a recipe for disaster.

Commanders who choose to perform hands-on tasks always seem to offer plausible explanations for why they do it. "We were short staffed." "The crew was inexperienced so I had to go with them." "It was an all-hands-working event." You can use any excuse you want to justify your actions so long as you understand you're impacting your command effectiveness, and the consequence can be significant.

The Comfortable Lure of Being Hands-On

At the beginning of the book, I told you about what happens to your body and brain when you are under stress. One of those things is your brain will direct your body to do those things you are most comfortable with (i.e., habits). So, can this impact a commander? In most situations commanders are promoted into their leadership positions after spending many years performing line duties like pulling hoselines, setting ladders, connecting to hydrants, pumping trucks, and operating rescue tools. It would probably not be a stretch to say that by the time line personnel get promoted to commanders, they have an expertise (and a comfort) with performing hands-on tasks at emergency scenes.

Under stress, these commanders may find themselves being drawn into performing hands-on tasks because that's what their brains are directing them to do—the tasks they are most comfortable with—the tasks they know they are good at. As I stated earlier, when this happens, commanders will offer very plausible explanations for why they "had to" perform hands-on tasks. Among the long list of excuses being offered is not likely to come: "My brain was instructing my body to do hands-on tasks because I'm more comfortable doing those activities than I am being the commander." A commander drawn into performing hands-on tasks can no longer maintain a big-picture view of the incident and this will, undoubtedly, impact his or her situational awareness.

One of my favorite parables to share is a tale of passengers boarding an airplane to go on a vacation. The plane takes off without incident and as the plane ascends, the captain's voice comes over the speaker system announcing:

> Good afternoon, ladies and gentlemen. On behalf of the entire flight crew I would like to thank you for flying with us today. And speaking of the entire flight crew, I should share with you that the copilot is out ill so I'm flying alone up here today. Oh, and the flight attendants who were scheduled to be on our flight were all reassigned to another one. However, don't you worry. When we reach our cruising altitude of 35,000 feet I'm going to engage the automated flight management system and come back into the cabin and serve you a variety of complimentary drinks and snacks. We also have beer and wine available for five dollars and, as always, exact change is appreciated. As I come through the cabin I will ask that those passengers sitting in the window seats in row one look out your windows for other approaching aircraft or fast approaching mountains. If you see either, please depress your flight attendant call button repeatedly. This will be my indication there is a problem. If this happens I will have about seven seconds to return to the flight deck, disengage the automated flight management system, and perform evasive maneuvers to avoid a collision with the other aircraft or the ground. Please help me keep the aisle clear by avoiding the use of the lavatories while I am serving the drinks and snacks. Thank you for your cooperation, and please let me know if there is anything I can do to make your flight more enjoyable.

If I were on that flight, I know what I would ask the captain to do that would surely make my flight more enjoyable . . . *Stay on the flight deck and fly the plane!* We'll either serve ourselves the drinks and snacks or do without them, thank you very much.

You're probably thinking that's a ridiculous story. That would *never* happen on a commercial airliner. You're right. It wouldn't. But as crazy as that parable is, a similar event can, and does, play out on emergency scenes. When commanders go hands-on, they are no longer in command because they've chosen to do something besides commanding the incident. In other words, they're serving the drinks and snacks. If I were at that scene, I think I'd be saying, "*Stay hands-off and in command.* There are other people to do the hands-on work. We need you flying this plane!"

In this chapter I shared several ways commanders can become overloaded and overwhelmed. I also identified some of the predictable behaviors of people who are under stress. From the perspective of situational awareness, being overloaded and overwhelmed effectively blocks your ability to develop and maintain situational awareness.

Responders may lose their ability to capture clues and cues (Level 1 situational awareness); they may no longer be able to make sense out of what is happening in the current moment (Level 2 situational awareness); and, they most definitely will not be able to predict the future (Level 3 situational awareness) because they can't even keep up with making sense of what is happening at the current point in time. This can lead to feelings of confusion, bewilderment, helplessness, frustration, and anger. This is not the condition you want to find yourself in as a responder.

Unfortunately, there's no pause button on an emergency scene. When a responder's workload exceeds capacity, it can be very difficult to get caught up. The situation seldom improves, because the incident complexity increases as time passes.

CHAPTER 10

HUMAN FACTORS

If you evaluate near-miss and casualty reports, interview expert-level commanders, or study other professions where high-risk and high-consequence decisions are made (e.g., aviation, military, surgery, air traffic control, and so on), you'll quickly understand there is a host of human factors that can impact situational awareness and decision making. You'll also notice the categories of human factors used in research studies and books vary greatly. When you hear the cause of a catastrophic event was "human factors," keep in mind the definitions and applications of human factors do not mean the same thing universally across disciplines. It's just one of those messy realities we have to contend with until the research disciplines agree on one standard definition.

Defining Human Factors

In my research and interviews with commanders, I found there were a variety of barriers impeding situational awareness that could be described as being *uniquely human*. Simply stated, there are some human traits that are tremendous assets to first responders. Conversely, there are some that can be categorized as liabilities, and these seem to carry a common theme. These can be physical, psychological, biological, or social in nature. I thus combined these to become my *human factors* category of situational awareness barriers.

Loss of focus

The first area I want to address is losing focus. I have talked with commanders who describe losing their focus in the middle of complex and stressful situations. Several described this loss of focus as being like an LCD projector that gets bumped, making the images on the screen no longer clear and sharp. Several factors can contribute to this barrier.

When you are under stress, your brain works really hard to make sense of what's going on around you, paying special attention to those clues and cues your brain can perceive as a threat to your safety. Through your hereditary instincts your brain has been exquisitely tuned to pay attention to stimuli that are moving, bright, loud, and in close proximity to you. In other words, you pay attention to the clues and cues in your environment that your brain perceives as the most threatening to your survival. As your brain focuses on those items, it can tune out some or all of the other clues and cues in your environment. Your brain simply perceives the nonthreatening clues and cues as less important. This may give the appearance that you're losing your focus because your brain is working hard to keep you focused on what it perceives to be life-threatening stimuli.

Also, you may recall from an earlier discussion in this book that your brain has a limited capacity to capture and comprehend data. When your brain is taking in fast moving and rapidly changing cues, it can stall, figuratively speaking, as it tries to make sense of the situation (figure 10–1). Or, it can drop cues as new stimuli come in before it's fully made sense of the old stimuli it captured just seconds previously. This may give you the sense that you are losing your focus.

Plug your ears

This phenomenon is not limited to visual stimuli. When you hear sounds, your brain often transforms the audible stimuli into visual images in your brain. For example, if a friend calls and tells you about her

Fig. 10–1. Stress in a dynamically changing environment can cause a commander to lose focus. When this happens, situational awareness will be impaired.

vacation, and she describes the white, sandy beaches, the pristine blue water, and the ice cold margaritas, your brain will form images of these things as a way to help you relate to what your friend is saying. You see what I mean? Even that last sentence is a testimony to the brain's capacity to form visual images from audible (or in this case, written) information.

The same thing can happen at an emergency scene. When a crew calls a supervisor on the radio and gives a report of what's happening, the supervisor's brain forms visual images based on the audible reports. It's possible for there to be so much audible stimuli (radio traffic or face to face), the supervisor's brain cannot keep up with processing all of it. When this happens, the brain can get overwhelmed and audible acuity (listening) can be impaired

Processing complex audible messages require the use of both the auditory cortex and the visual cortex. The message enters through the ears; the hammer, anvil and stirrup send the vibrations to the auditory cortex; and the message is converted into electrical impulses—sort of like a cognitive Morse code. The coded message is then sent off to the visual processor to be formed into a visual image of the message in the brain. If the visual processor is busy, the audible message doesn't get processed. This can be experienced in two forms: as a complete lack of awareness of the message (i.e., auditory exclusion), or as a distortion of the message (sounding sort of like an FM radio slightly out off channel, the message is "blurry" or out of focus in the context of this discussion).

It's important to understand that the loss of focus, in the context of this human factors discussion, doesn't happen on purpose. You can, however, miss visual and audible cues because you are being complacent (e.g., letting your guard down or being lazy), which is discussed in another chapter. As a human factors barrier, responders would not consciously choose to have blurred vision and distorted hearing. Nonetheless, it's a physical phenomenon of which you need to be aware.

Role ambiguity

Role ambiguity at an incident and the lack of clearly defined and communicated operational directives, strategies, and tactics can also contribute to a loss of focus. Someone needs to ensure there is a common understanding of purpose, roles, and goals. Otherwise, in a dynamically changing environment replete with complex stimuli, it can be easy to have your attention (focus) drift from one thing to another.

As I think about this concept, I am reminded of a story about a couple who were riding home in the car from church one Sunday. The wife made a comment to the husband about a particularly hideous outfit a lady had worn to church, to which the husband replied that he had not noticed the outfit. Then the wife commented about how misbehaved the children were in an adjacent pew. Again, the husband replied that he had not noticed. Then the wife commented about a single man who seemed to conveniently sit beside a different single woman every Sunday. Again, the husband replied that he had not noticed. Finally, in her frustration the wife snapped back at her husband, "Why do you even go to church if you're not going to pay attention?"

In this example, each person could make an argument that the other had lost their focus because each had separate goals (incompatible as they were) for being in church. Maybe stated another way, each saw what they were looking for. It would be reasonable to assume the woman did not get much from the sermons because she was so focused on her physical surroundings.

It's not always obvious when you lose your focus. As in the church story, the wife was paying attention to what she thought was important, and so was the husband (though the story doesn't explicitly say where his attention was focused, it could be assumed he was actually listening to the sermon).

You know what happens when you assume?

I recall my high school science teacher asking this question to the class following a student saying he had made an erroneous assumption about when an assignment was due. After the question was asked, there was a period of silence. Then the teacher answered the question, "You make an ASS out of U and ME."

Because first responders rarely operate in an environment where there is complete and accurate information, a certain amount of assumption is necessary if timely decisions are to be made. However, poor assumptions can lead to bad outcomes and, unfortunately, poor assumptions can be another human factors barrier to situational awareness. As noted, at emergency scenes it is an essential requirement that assumptions be made because there is a lack of complete information, and the information that is available changes rapidly. Thus, a responder is forced to make certain assumptions to help fill in the unknowns.

Assumptions are based, in part, on your emergency services education, your training, and your first responder experiences. They are influenced by many other factors, too, including your biases, religious beliefs, morals, upbringing, and friends and acquaintances—essentially all of your life's experiences. Assumptions help form your beliefs and help you explain (or make sense of) your experiences even in the absence of supporting factual data. Making assumptions at emergency scenes and speeding up the decision-making process can be an asset to responders—IF your assumptions are correct.

The outcome can be less than desirable if your assumptions are incorrect. In some ways an assumption might be compared to a scientific hypothesis, except many assumptions occur in the subconscious brain and outside your conscious awareness.

Simply stated, it's a protocol that goes something like this: Make an assumption that if you try something you will get an expected outcome. Try it and see if it works. If it does, gather the facts and note the outcome so you can replicate the results the next time. If it doesn't work, gather the facts and note the outcome so you can avoid the undesirable outcome. Through this system of trials and outcomes we are learning (hopefully).

It's important for responders to seek facts that validate their assumptions. Think of this as a mental validation of the accuracy of assumptions prior to committing to an action—facts that have been present (or absent) from previous experimentation (learning) that support or refute assumptions. Responders can get into trouble quickly when they make assumptions and then overlook the step of fact finding—an essential component toward validating the accuracy of assumptions.

Your assumptions can be strongly influenced by your beliefs and biases. This is especially true in the absence of strong evidence. An unrelated example is the debate on whether humankind evolved from primates or were divinely created by a higher power. This issue speaks to the core of a person's belief system and his or her knowledge (and belief) of scientific theories of evolution.

The root word of assumption is "assume"—a word whose definition includes references to one's belief that something is true or accurate without verification or evidence.

Your assumptions are derived from your collective life experiences. And, as much as you might disagree, you don't know everything. Yet you may hold a strong opinion about something even though you have little evidence to verify or validate your belief. Caution is in order when making assumptions on emergency scenes. Seek to validate assumptions with some amount of data that supports or refutes them. In his book on leadership, Colin Powell spoke about his method for gathering facts and data in the process of making military decisions. There he noted his goal to gather 40 to 70 percent of the facts and then to draw on his intuition and experience to fill in the rest and make his decision. That seems like a pretty good rule of thumb for first responders too.

Confidence and Ego

Confidence and ego can also be human factors barriers to situational awareness, and both can impact decision making. These barriers can be tricky because, together, confidence and ego are very important qualities for first responders. As previously noted, responders rarely have complete and accurate information when making decisions; therefore, it's essential to make assumptions. This requires confidence in one's knowledge, skills, and abilities. However, excessive confidence and an inflated ego can adversely affect the quality of decision making.

When responders are on the scene of an incident they have experienced many times before, they can suffer from a bias of confidence—a false sense that this incident is like all the others where circumstances have been similar. Similar—but not exact. And this is where over-confidence becomes a situational awareness barrier.

I knew before I arrived

Responders may pre-load facts and data into their mental model (i.e., schema) based on previous experiences, not the current emergency scene. This pre-loading may occur even before responders arrive on the scene of the emergency. When they arrive, the responders may not develop a situational awareness using the actual clues and cues present; rather, their situational awareness was formed based on *assumed* facts and data, driven by over-confidence, derived from previous experiences, and all done prior to arrival at the emergency scene.

Confidence is a balancing act

Lest you think that having confidence is bad, let me assure you that it's not. In fact, confidence is essential to making quality decisions under stress. Absent complete facts, responders must have the confidence that the information they have been able to gather during their brief and rapid size-up is accurate and complete enough to formulate an action plan that will not get someone hurt or killed. If responders lack self-confidence, they may delay making critical decisions while waiting for more information—information that may never come. If responders are over-confident, they may make decisions based on their assumptions without validation—and they may be completely wrong.

Ego

Ego can also have an adverse impact on first responder situational awareness and decision making. This can be witnessed during scenarios where a responder conducts a size-up, makes a plan, and puts the plan into action, yet the plan was either flawed or the conditions changed, rendering the plan obsolete. Responders with inflated egos may find it difficult to admit mistakes, concede defeat, or acknowledge their failures. Thus, they may stay the course on an action plan that is not headed toward a positive outcome. Even though they know and can see that their plan is not working, they convince themselves to stick to it—just a few more minutes . . . and then a few more minutes . . . and then a few more minutes. All the while, conditions are changing, and the situation is getting worse. Some responders with inflated egos may even become more aggressive and deploy more resources when their plan is not working (versus retreating). Ironically, in their effort to avoid acknowledgment of their failed plan and their insistence on trudging onward, they lead the incident directly into an outcome that results in a catastrophic injury or loss of responders.

I'm reminded of a story I heard about a guy who was, for all intents and purposes, a normal guy. But he went on a trip with some friends to a gambling destination. While on the way there he bragged to his buddies about how good he was at playing blackjack (though he'd never played it other than a gaming version he had on his home computer). He built himself up to the point that he became the envy of his friends. Needless to say, when they arrived at the casino, his ego was inflated and he had a lot to live up to.

Unfortunately, the cards were not being dealt in his favor, and he lost more than $100 in his first half hour. Now, a logical well-grounded person with a balanced ego might think to himself, "I'm not doing so well today. My luck is down. My strategy isn't working. Perhaps I should just leave, try another game, or at least change my strategy." Unfortunately, he did none of the above. Instead, he kept playing the same game, the same way, with the same result. Even more unfortunate was the casino's willingness to extend him a line of credit (give him more resources to throw at his problem). So he kept betting more and more and more, hoping and praying his luck would turn around and his strategy would work. You might say he was fulfilling the definition of insanity—by doing the same thing over and over and over again, expecting the results to be different. By the time his friends caught up to him, he'd lost almost $2,000 (figure 10–2).

Fig. 10–2. Gamblers who are down on their luck may up the ante to try to recoup their losses. The same thing may happen if an incident action plan is not working. Commanders may press their luck and increase their risk taking.

The gambler had let his ego affect his ability to make good decisions. He stayed true to a plan that was not working. He could not bring himself to admit his defeat and go in a new direction. Thankfully no one got hurt or killed as a result of his decision making (though I don't know what happened to him when he got home and had to explain his losses to his wife).

It's human nature for a person to desire respect and to be liked. However, for some people, this trait is not in balance. Out there in the world, and unfortunately in some emergency services organizations too, are people who think they're smarter than everyone else and who feel they are always right. For people with this affliction, it may, quite literally, be impossible for them to admit when they were wrong (because in their minds, they're never wrong).

Addicted to being right

I recently read a study where scientists were able to isolate a chemical in the brain that is released when a test subject answered a question correctly. The neurological response was the release of a chemical that gives the person a mini "rush"—a jolt of feel-good juice, if you will. Some researchers have hypothesized people can become addicted to this chemical, and thus, must always feel they are right in order to get the rush.

Research has taught us that mind-altering chemicals are released when a person falls in love or becomes angry, jealous, happy, or depressed (as well as a host of other emotions). Science supports the notion of a chemical release tied to the feeling of being right, and that one could become addicted to the rush that comes from believing they are correct. It certainly helps to explain the behavior many of us have witnessed in some people we've come across throughout life—people who are blissfully happy in their self-proclaimed ability to *always* be right.

A confident commander can be good. An under-confident commander can be dangerous. An overconfident commander can be deadly.

Is Perception Reality?

Another human factor that affects situational awareness is having a flawed perception of reality. Everyone has perceptions of reality. For all individuals, their perception is their reality. However, simply because they perceive something as real does not make it so. Filters that allow people to see what they want to see and avoid seeing things they don't want to see influence perceptions.

Here's an exercise you can do that demonstrates, in a clear and vivid way, the impact of our filters on our perceptions of reality. This is an exercise I devised in an unexpected way. I had received a pair of yellow-tinted driving glasses as a gift. When I put them on I realized they changed the color of some things and neutralized the colors of others. Armed with that knowledge I went to an office supply store and bought some paper of various colors. When I put the glasses on, it distorted the true color of the paper.

Thus, I had the makings of a *perception versus reality* experiment. When I start this demonstration I ask each participant to put on the yellow-lensed glasses as I hold up pieces of paper of various colors. I then ask the participant to announce aloud what color the paper is. As the participant announces the color, I write that color on the piece of paper with a large black marker, and I tape the sheet of paper on the wall. We do this with four or five pieces of different colored paper. Then I ask the participant to remove the glasses and tell the rest of the group the color of each piece of paper. It's evident the participant's perceptions were off. The green piece has "blue" written on it. The yellow piece has "white" written on it. And the blue piece has "green" written on it.

During the exercise the participants stated the colors as they saw them. The glasses are the equivalent of the filters of lived experiences, biases, and beliefs that every person applies (unknowingly) to every situation they evaluate. If it weren't so obvious that the participant was wearing the glasses that skewed the perception of reality, it would be easy to see how a disagreement could ensue as one person says the piece of paper is yellow while another states confidently that it's white.

At the scene of emergencies responders also have various perceptions that can lead to differences of views of what is happening and, thus, can form very different opinion/beliefs. Unfortunately, many times when there are differences of opinion on emergency scenes the person in the position of formal authority (e.g., the supervisor) is probably going to win. In reality, the supervisor may very well be the one with the flawed perception of reality. Rank does not make a supervisor have better situational awareness. It just gives the supervisor the authority to declare him- or herself right.

Culture

Another human factor trait, although not individual, that can impact situational awareness is the organizational culture. Culture can be a significant barrier to good decision making. Emergency service organizations, and individuals that comprise those organizations, are rich in traditions and pride. Stories abound of aggressive tactics resulting in lives and property saved resulting in firefighters being elevated to *hero* status for having successfully performed acts of bravery—some perceived, some real. In some organizations, the competition among companies and members to one-up each other can be fierce.

I was told of one department that had a special rescue unit that responded to every working structure fire. This unit was staffed with the department's *elite* members—the self-anointed *best of the best*. When this squad arrived, their mission was to extinguish the fire and rescue occupants. Their operators drove ridiculously fast and recklessly in an effort to ensure the squad's early arrival at the scene and reputation for first-water-on-the-fire status. If the squad did not arrive first, regardless of what was already deployed, the squad's crew would advance its own attack line into the building. If needed, they would quite literally climb right over the backs of the attack crew if they are not advancing their line fast enough to suit the rescue crew (figure 10–3).

Fig. 10–3. In organizations that have adopted a culture of safety, crews know when conditions are right to be offensive-aggressive; when they are not, the tactics should be defensive-conservative.

This story came to me as a first-hand account from a person to whom this happened. He has since left the department and accepted a career position with another department whose culture (as he describes it) is far less reckless and dangerous.

Culture can impact the department's decision making in several ways. For example, if a department's philosophy is one where the members believe their reputation is to be an aggressive, interior structural firefighting department when they arrive on the scene of a working fire their strategy may *always* be to go offensive, aggressive and interior, every time, regardless of the situation found. Obviously this can be dangerous because not every fire is suited for an aggressive interior attack strategy.

An equally dangerous condition can arise from failing to know when to back out—or transition from offensive to defensive attack strategy if the fire is not contained in a reasonable amount of time. An organization's culture may shun—perhaps even ostracize—a member for even suggesting a strategy to start defensive or go defensive after the offensive attack has been initiated.

The powerful need to be liked

This environment can have an adverse impact on decision making. Chances are supervisors know the culture because they were raised in it, participated in it, and perhaps even fostered it. In such a scenario, supervisors might arrive at an incident where the crews should be ordered to take a defensive posture, but the order is not issued. Instead, the supervisor lets the crews engage in an interior attack even though the supervisor knows better. Why would a supervisor allow an interior attack in a situation known to be unsafe?

That question was among the research for which I sought answers, and I was surprised with what I learned. Some of the supervisors I interviewed described this very scenario to me and admitted they allowed (or did not stand in the way of) aggressive, offensive attacks where they should have never been allowed to occur. I'm told their organizational culture rewards and celebrates those who embrace highly aggressive tactics. Those who opposed the tactics, regardless of the situation, are ridiculed. The supervisors confided they didn't want to risk being ridiculed by their subordinates. Being popular with the crews was more important to them than making the decision to be defensive.

> *One supervisor shared that on multiple occasions he let crews go in when he should not have because, as he stated, "I knew that if I didn't, when we got back to the station, they would kill me and eat me."*

Another supervisor emphatically stated that he would never cave in to the pressure of their organization's culture, and if he determined the incident should be defensive that is what he would order it to be. This strong command presence is to be applauded. This supervisor's mind-set contributes to the prevention of near-misses and line-of-duty deaths. Tragedy is far less likely to strike when a supervisor makes the right call and stands up to the pressures of the organization's culture.

During another discussion on the pressures of organizational culture, a supervisor stated,

> *"It's not my job to be popular. It's my job to keep my employees safe. And if they don't like my decisions that ensure their safety, I'm not going to lose any sleep over it. They can hate me for the balance of their long and healthy lives."*

An aggressive-minded supervisor may fail to see the door has closed on the opportunity to save a life (or property). When a supervisor orders the crew to engage in an offensive attack, the subordinate firefighters may follow the direct orders of their superior officers without question.

Should a first responder have the right (or obligation) to question the orders of a superior (in a respectful manner) if the subordinate believes the superior officer may be making a critical life safety error in judgment?

When I ask this question to responders, it typically triggers a debate because many supervisors do not favor the idea of having an insubordinate underling suggesting they may be wrong at an emergency scene. The problem can be compounded because some line personnel possess a high level of respect and loyalty for their superiors and would never question orders.

At a meeting with the members of a large municipal fire department, the topic of organizational culture and how to point out when a superior officer may be overlooking an important safety concern came up. One participant sat bolt upright in his chair and confidently barked, "My role as a firefighter is to follow the direct orders of superior officers without question."

The problem with this mind-set is after you strip off all the gold bugles and titles, supervisors are human beings—*fallible* human beings. They are susceptible to all the barriers discussed in this book. They can miss crucial information during their size-ups. They can misunderstand radio traffic. They can have flawed assumptions and perceptions. Supervisors who consider themselves flawless are as dangerous as subordinates who think their supervisors are above making mistakes and thus should be blindly followed.

Tell me if I'm going to step in dog poop

I believe all supervisors (sans those whose egos are out of balance) want to be told, respectfully, if they are overlooking something important that could save a responder's life (figure 10–4). I truly believe no healthy-minded supervisor wants to live with the consequence and regret of a bad decision that could have been avoided if only someone would have simply spoken up.

Fig. 10–4. Competent commanders are approachable and appreciate being informed (respectfully) if they are making a mistake or overlooking something important. At the end of the day, the goal is for everyone to go home!

Complacency—letting your guard down

Another human factors barrier that can impact situational awareness is complacency—the feeling of satisfaction and low vigilance people may experience when they become at ease with a situation. The more routine or mundane an emergency call may appear, the greater the potential for responders to become complacent. This is especially true for experienced responders who have seen similar incidents over and over again. They can get lulled into a false sense of security, believing this incident is just like the last one and the one previous to that. When responders become complacent, they can let their guard down. This may cause them to stop capturing clues and cues (flawed Level 1 situational awareness), and they may miss important visual or audible indicators that this incident isn't like the last one.

While conducting my research I interviewed one commander who shared with me the details of an incident where the fire was a "routine chimney fire" in a two-story single-family residential dwelling.

When the commander arrived, the occupants were out of the house and there was nothing showing from the street level. The commander said he breathed a sigh of relief and then he became complacent—he let his guard down—because he'd responded to hundreds of chimney fires throughout his career. As the crews arrived, he ordered them to the interior. One crew opened an attic hatch and reported smoke and some fire in the attic. The commander ordered a hoseline to the attic hatch and instructed another crew to open the roof to relieve the heat and gasses. The interior crew reported the fire knocked down, and roof crew reported the hole had been cut and the attic was venting.

The interior crew reported water was leaking through the ceiling following the fire attack. The commander ordered a couple of companies to take some salvage tarps in and cover the belongings in the family room. "It was a routine operation," he assured me. The commander then got side tracked talking to the homeowner and a few other officers standing around his command vehicle. He was no longer monitoring the situation. The fire in the attic flared up a few times, but the nozzle person was right there to keep it in check. Then, without warning, "Mayday, Mayday, Mayday!" came the scream over the radio. "We're trapped in a collapse on the second floor." The commander was very confused. He was definitely caught off guard because of his complacency with the incident. He didn't know how many firefighters he had inside at the time of the Mayday or where they were.

Note: The water was being applied in the attic by a newer firefighter who was enjoying his nozzle time. The massive amounts of water he flowed was being absorbed by the attic insulation and sheet rock. Finally, the weight of the water was more than the ceiling could hold, and the entire ceiling collapsed, trapping and injuring six firefighters.

This commander found himself in a deficit position—the kind of position that no commander ever wants to be found in—not knowing what's happening at the incident because of complacency (figure 10–5).

Fig. 10–5. Even when everything appears to be going as planned, a competent commander must remain vigilant for changing clues and cues, some of which can be subtle and slow to change.

Here's a Dollar . . . Now Pay Attention!

A responder can also be affected by simply not paying attention. Unfortunately, not everything that changes at an emergency scene happens in an obvious way. Some changes are very slow and, in fact, so subtle that if you're not concentrating and looking for them you may miss them. Some things change so slowly that they're hard to see even when you are looking for them.

To have Level 1 situational awareness (the ability to capture the clues and cues of the things that are happening around you), it is essential that you be conscious, present, and scanning your environment for subtle changes.

I have tested attentional abilities numerous times in exercises using first responders. Without exception, the majority of the participants fail to see when changes happen before their very eyes. It's disturbing because when I'm doing my exercises we're in a controlled environment where there is no stress and the conditions are ideal. If you purposely take your eyes (or ears) off the incident, you can miss important clues and cues in a matter of seconds.

Paying attention is not as easy as many people think, and situational awareness is so much more than simply *paying attention*.

CHAPTER 11

ATTENTION MANAGEMENT

While operating at an emergency scene, there is a lot of activity that competes for a responder's attention. Some believe the essence of situational awareness lies in one's ability to pay attention. To some extent, that's true. One element of situational awareness is paying attention. Previously, I discussed how situational awareness is developed in three levels. The first level is capturing the clues and cues in each situation you find yourself in. If you are not paying attention, it's difficult to capture both audible and visual cues and clues. (Note: It can also include clues and cues you taste, smell, and feel.) Unfortunately, what to pay attention to is not always obvious or intuitive, especially while working in a dynamically changing, stimuli-rich environment.

Hereditary Instincts

Hereditary instincts can be both helpful and hurtful when it comes to being attentive. Recall from our earlier discussion that when a person is in a high-stress, high-consequence environment, the body's fight-or-flight response is triggered. In this condition, the multiple sensory systems (e.g., visual, auditory, tactile) instinctively go on high alert. This causes the eyes and ears to become hypervigilant (highly sensitive) for anything the brain may interpret as a threat.

Instinctively your attention may be drawn to objects that are bright, loud, fast moving, and in close proximity to you. This can be beneficial when the things you need to pay attention to are bright, loud, fast moving, or in close proximity to you. But what about when the important clues and cues that ensure your safety are dull, quiet, still, or far away? Responders need to be especially alert for these subtle indicators because your eyes and ears are not naturally attuned to looking for or listening to them under stress.

Adding to the challenge of seeing, hearing, and processing subtle cues and clues are the distractions and interruptions a responder can endure while operating at an emergency scene. When I conducted my research, I classified distractions and interruptions separately. Whereas the impact and consequences may be similar, the means by which your attention is drawn away can vary.

> *Distractions inadvertently draw your attention away from what you are supposed to be watching or listening to.*

For example, if while operating at an emergency scene there's a loud noise behind you (e.g., a siren, air horn, scream, or explosion) you may turn to look in the direction of the noise and focus your attention on figuring out what's going on—that would be a distraction. Stated another way, distractions draw your attention away *accidentally*.

> *Interruptions, on the other hand, draw your attention away on purpose.*

Say, for example, while operating at an emergency scene someone approaches you for the purpose of having a conversation or to share some information. This causes an interruption to your thought patterns, and it can draw your attention away from seeing (and hearing) other important clues and cues happening at the emergency. Interruptions are *intentional* whereas distractions happen *unintentionally*.

Regardless of whether you are drawn off task by an interruption or a distraction, the consequences can be significant. Drawing a responder off task impacts Level 1 situational awareness and, as a result, the responder may be challenged to capture important clues and cues while trying to overcome the distraction or interruption.

Selective Listening

Responders can also be affected by selective listening. This barrier to situational awareness can manifest itself in several different ways. For example, someone can approach you to have a conversation (as in the interruption example noted previously). If you're a supervisor or commander, hopefully you're listening to the radio communications of your most at-risk personnel operating on the emergency scene or, depending on the complexity of the incident, at least listening to the radio traffic of the division or sector supervisors who are overseeing the most at-risk operations.

When a person approaches you, it can trigger a hereditary instinct and you may perceive it as a threat. This may be enough to put your brain on high alert for danger because that moving object (in this example it may be your operations officer coming to have a face-to-face discussion) represents a danger to your primal brain. It doesn't matter if the person approaching you is someone you know well and don't fear. Hereditary instincts triggered in your brain see moving objects in close proximity as potentially harmful and will alert your senses to pay attention. This includes your eyes watching and your ears listening. Anyone approaching you is going to garner your attention, and you may find yourself selectively listening to that person to the detriment of other important cues (like radio traffic from at-risk personnel). You may also find yourself paying attention to that person to the detriment of other important happenings on the incident scene.

During my research interviews, some responders even acknowledged putting their hand over the speaker on their lapel microphone, holding their radio up against their chest to muffle it, or even turning the volume control on the radio down so they could carry on a conversation with someone face to face.

It's not for me to judge whether the face-to-face conversation is more important than the radio conversation; however, the crew(s) operating in the high-risk environments may not be in a position to repeat emergency radio traffic a second time if missed. Conversely, the person you are having a face-to-face conversation with is in a much better position to repeat their message if necessary.

You Can't Multitask Listening

As discussed in the previous section, humans can struggle to effectively listen to and comprehend multiple verbal conversations simultaneously. This underscores a critically important associated lesson about your situational awareness. Responders who try to listen to multiple conversations simultaneously are very likely to miss clues and cues or not comprehend the meaning of the messages (figure 11–1).

The act of listening requires conscious thought as your brain processes the audible messages into visual images. Notwithstanding the amazing abilities of the brain—unrivaled by any computer ever invented—it does have limitations. One limitation is being able to process and comprehend only one conversation at a time. It's possible for you to bounce back and forth between multiple conversations and derive the general idea of each conversation. But something is sure to be missed as you listen to one conversation to the detriment of the other.

Fig. 11–1. Effective commanders will know they can only listen to and comprehend one radio channel at a time. A radio set to scan or attempting to monitor multiple radio channels may cause critical clues and cues to be missed, impairing situational awareness.

Trying to listen to two conversations is like trying to watch two shows at the same time on two television channels. You can flip back and forth between the channels as often and as fast as you want. No matter how good you are at it, you're going to miss something.

Narrowing of Attention

Another way attention management can be impacted is by the narrowing of your attention. This concept is often taught in basic emergency responder training programs as *tunnel vision*. This biological reaction in the brain, triggered by hormones and chemicals, causes your senses to narrowly focus on stimuli that are perceived as threatening. To some extent, the narrowing of attention at an emergency scene can be a good thing. For example, you would not want to be at an emergency scene and thinking about the leaky roof on your house or your broken-down car. You want your attention—all of your attention—to be focused on the emergency. In fact, this is an essential component in the development of Level 1 situational awareness—being focused in the most important things happening in your environment.

However, when you're under stress your attention can narrow down even further to one geographic area of the incident scene or to one task or operation being performed on the incident scene. This is covered in chapter 6 where I discuss how understaffing can affect a supervisor's situational awareness. Recall that when a supervisor deploys an understaffed crew into a high-risk operation, it creates stress. This may cause the supervisor's attention to narrow down to that understaffed crew, where they are, and what they are doing. When this happens, the supervisor can miss other really important visual and audible clues and cues.

A responder can also suffer from task fixation. Becoming fixated on the performance of a single task at the emergency scene can cause the responder to become oblivious to other events. For example, an incident commander and the first engine company arrive together to find a victim trapped on the second floor. The crew is sent in to perform a rescue. The commander and/or the company officer may fixate their entire attention on the performance of the rescue. This may cause either of them to overlook other very important tasks that need to be performed, such as completing a 360-degree size-up, deploying an attack line, ventilating, or securing a water supply.

One scenario shared by a supervisor during the interviews for my research serves as a great example of task fixation. The pump operator couldn't get water flowing from the pump for one reason or another. (Note: When this happens, it's rarely a mechanical issue with the apparatus pump, but most often an operator error induced by lack of realistic and repetitive training on a task that is being performed under stress.) When the crew is calling for water and it's not coming, the stress levels of everyone can go up. A commander who hears the desperate calls for water may feel compelled to physically go to the apparatus and try to resolve the pump issue. This may take only a few moments, or it may (depending on the experience level of the commander as a pump operator) take quite a while to troubleshoot. All the while, the commander's attention is fixated on the single task of getting water out of the engine and is no longer focused on the big-picture incident. This may cause the commander to miss critical visual and audible cues and clues that can affect situational awareness (figure 11–2).

Fig. 11–2. When an incident commander's attention is drawn to a specific task—such as helping a struggling apparatus operator—he or she can miss critical incident clues and cues. This can quickly impair the commander's situational awareness.

A powerful (and sad) example of how situational awareness can be impacted by task fixation comes from aviation and the doomed Eastern Air Lines Flight 401 in 1972. As the flight was preparing to land in Miami, the crew flipped the switch to lower the landing gear (wheels). If the landing gear goes down properly, a green indicator light illuminates. The light didn't come on, so the crew aborted the landing and went up to a safe altitude, set the automated flight management system (the autopilot), and started the process of assessing and resolving the problem. Both the captain and the copilot were looking through their flight manuals to figure out how to fix the problem. One of the suggested fixes in the manual was to replace the light bulb with another bulb from the console that was confirmed to be working.

Unfortunately, the lens cover over the landing gear indicator bulb was cross-threaded and not cooperating. The pilot recalled he had a small screwdriver in his satchel that he thought he could use to pry off the cover. As he reached around to his get the tool, he unknowingly bumped the yoke, which disengaged the autopilot. As the pilot and copilot continued to work at getting the cover pried off and the bulb swapped, they became fixated on that single task. Neither pilot noticed the autopilot was disengaged and the plane was losing altitude.

Fortunately, the air traffic controller noticed the altitude deviation and called Flight 401 to check up on them. The pilot radioed back that they were working on resolving the problem and would soon be in a position to make another landing approach. This transmission led the air traffic controller to believe the pilots were aware of their declining altitude. Soon after, a warning system activated to alert the pilots they were dangerously close to the ground. Unfortunately, by the time the pilots were able to "un-fixate" from the task they were working on (the burned-out light bulb), the plane crashed into the Florida Everglades. The investigation report revealed the landing gear lowered properly in the first place. The pilots were so fixated on a single task that neither one was flying the plane. The only problem with the plane was a burned-out light bulb. It's absolutely essential that someone look at the big-picture incident at all times (figure 11–3).

Fig. 11–3. A good situational awareness best practice that can help a commander avoid task fixation is to assign other command-level officers to sectors and divisions, allowing them to focus on task-level work and report their progress back to the commander.

Aviation learned a painful and costly lesson that evening. No one was flying the plane. Now, every emergency checklist used in commercial aviation has, as the first step, "Fly the aircraft." You would think that pilots would have enough training and expertise to rationally know someone needs to fly the aircraft. Unfortunately, doomed Flight 401 proved otherwise. Under stress, you can become fixated on a single task and, as a result, miss other important clues and cues.

CHAPTER 12

COMMAND SUPPORT

Command support is a barrier that has a direct connection to staffing levels. Emergency scene workloads can be intense and can cause commanders to quickly become overwhelmed. Some organizations have the benefit of having command aides assigned. Others have had such positions cut. Some have never benefited from having assigned command support. This is unfortunate, because the value of having this support can greatly improve situational awareness and incident safety.

Sharing the Load

One of the most valuable lessons I derived from my research was gaining an intense understanding of the complexity of managing an incident and the impact that stress, attention management, and workload management has on incident commanders and company officers. I am completely and thoroughly convinced that commanding a complex emergency incident is not a one-person job. There are dozens of examples shared throughout this book that demonstrate where a responder's situational awareness was impacted by one or more barriers, and the impact could have been significantly reduced if the responder's mental workload had been shared with another person. This person could be assigned to do a walk-around, monitor the radio for a supervisor, complete the worksheet for the supervisor, use a checklist to ensure the supervisor is doing everything properly, and run interference for the supervisor from people who want to have face-to-face conversations. My research revealed example after example where barriers to situational awareness could have been reduced if the responder had the benefit of a second person to serve as an aide, advisor, or assistant.

In aviation, a commercial airliner would never fly without the benefit of a co-pilot and, in some cases, a third officer on the flight deck. Why? Because the stress and mental workload is too much for one person to handle, especially if something goes wrong. The same stress and mental workload befalls emergency incident supervisors at complex incident scenes. Simply stated, there's too much for one person to handle—too many visual cues, too many verbal cues, too much technical information to process, too many tasks to keep track of, and too many personnel to account for.

When staffing levels cause an incident to be managed by a single, overloaded supervisor, it's a setup for failure. I'm sorry to be so blunt about it, but it's a fact. The human brain's limited ability to process and comprehend information, compounded by stress, creates limits on cognitive performance.

It's a commonly accepted practice for emergency responders to enter high-hazard environments in teams of two, three, four, or more (figure 12–1). These responders know they should never work in these environments alone; it's too risky. The physical and mental demands are too great; it's a buddy system.

> *You look out for me and I'll look out for you. If something happens to you, I'll be there to help you out. If something happens to me, you'll be there to help me out. If we need to perform a rescue, we'll accomplish this physically demanding task by working together.*

Fig. 12–1. Firefighters routinely work in teams to share the workload and to look out for each other. This concept needs to be extended to incident command positions as well. The cognitive demands on a commander can easily exceed the capacity of one person.

When it comes to frontline, hands-on duties, the concept of teamwork is a commonly accepted practice; however, when it comes to commanding incidents, there's a fundamental flaw. Budget and subsequent staffing reductions have created situations where commanders routinely and regularly respond and work alone. Whereas some departments have retained drivers for their command officers, it's been opined in many municipalities that a command aide (who may also serve as the driver) is a luxury and many of these positions have fallen under the budget axe. From the perspective of commander situational awareness, this is a tragic consequence of financial cutbacks. Science readily supports the notion that the cognitive demands of an incident, where multiple companies are working, exceed the mental capacity of one person. Command an incident with less than adequate cognitive horsepower and the risk of a bad outcome from flawed situational awareness increases significantly.

Think about the start of an incident as the point when the alert tones are transmitted and units are responding. There's usually a lot of radio traffic from both the apparatus and dispatchers sharing caller information. Commanders being chauffeured to the incident are able to dedicate their complete cognitive attention to this information (just like a company officer can who is riding in the right front seat of an apparatus).

When commanders must drive themselves to the scene of the emergency, the mere act of driving consumes some cognitive capacity. The act of driving can also increase the commander's stress level (especially in congested traffic situations). As the commander focuses on the act of driving, important clues and cues being transmitted about the emergency may be missed (figure 12–2). Stated another way, when commanders must drive to the emergency, it can compound the impact on their situational awareness, as they may also simultaneously be listening to their radio, talking on their radio, talking on their cell phone or two-way direct-connect telephone, making notes about the dispatch information and responding units, and they may even be trying to access facility information on a mobile data computer. It's not difficult to see how commanders could become overloaded and overwhelmed in this situation. And they haven't even arrived on the scene yet.

When at the scene, a commander (who may be the first-arriving company officer) working alone continues to be challenged by the management of high volumes of information, rapidly changing conditions, a physical scene that may be impossible to see from one location, high volumes of radio traffic, numerous units arriving that need deployment assignments, and occupants and witnesses who may have critical information to share that may be vital to responder or victim safety and survival.

Where response procedures are such that commanders routinely arrive and work alone in the performance of their duties, I contend the solo commander, in a high-stress, fast-moving, high-consequence environment, may very well be set up to fail. No one should be surprised when this scenario results in a near-miss or casualty incident. Critics will quickly line up, point their fingers at commanders, and say they should have done this or they should not have done that. It's easy to be a critic on the proverbial Monday morning, sitting around a table in a no-stress environment. Commanders do not become overloaded and overwhelmed

Fig. 12–2. Having a driver for a command officer is viewed in some systems as a luxury. That is unfortunate because the mental demands on a commander start well in advance of arriving on the scene of an emergency. When commanders must drive themselves to emergencies, a significant amount of their cognitive capacity is going to be consumed by the act of driving.

on purpose, and they certainly do not lose their situational awareness and make bad decisions on purpose. In some cases they are simply victims of running a very complex incident solo and they're overwhelmed.

I don't want to diminish the fact that there are responders who are undertrained, inexperienced, and underprepared to command complex incidents. Responders who lack training and experience are even more vulnerable to be overloaded and overwhelmed as their brain struggles even more to make sense of events at such incidents. The more new or novel an event is, the more mental capacity it takes to sort out the clues and cues to figure out what is going on and how to manage through it.

When responders are put into a position that requires them to see, hear, and process many complex pieces of information without the proper help, it's tantamount to an automatic setup for failure. That's why there are surgical teams in a hospital surgery suite and a minimum of two pilots on a commercial airline flight deck. There's a reason for teams of people in high-stress, cognitive-demanding environments; they share the workload. But in emergency service organizations, quite often supervisors have no support team—they operate alone. They can easily become overwhelmed trying to capture the clues and cues (Level 1 situational awareness), comprehend them into meaning (Level 2 situational awareness), and make accurate predictions about where the incident is heading (Level 3 situational awareness). Such a situation can cause the supervisors to become frustrated and angry, which triggers even more barriers to situational awareness—all because they don't have someone to share the mental workload.

Checklists and Worksheets

There are some tools responders can use to help manage memory and workload, namely checklists and worksheets. However, some responders admit they don't use checklists, worksheets, or command boards as often as they should. There are several reasons for this. Some simply didn't have them; others admit they have them but forget to use them; still others acknowledge they get so busy they forget to keep them updated—perhaps another symptom of work overload. Some acknowledge they never got around to getting the command board out of the trunk of their vehicle. In a few instances responders noted their command boards are massive in size and too cumbersome to set up. So they reason they simply didn't have time for all that assembly.

Using checklists and worksheets can be very helpful. They help manage memory, and your memory is particularly vulnerable to degradation under stress. As it relates to memory, you essentially have two types that you want to manage. First is the memory of everything you have already done. This is known as *retrospective* memory. Second is the memory of everything you have yet to do. This is known as *prospective* memory. Worksheets help you manage retrospective memory because you can write down and keep track of all that's happened, where your personnel are located, and what tasks they are performing. Checklists are helpful in managing your prospective memory because they contain check boxes of the important tasks you need to do at the emergency scene but have not yet done (figure 12–3). Of the two types of memory, your prospective memory is the more fragile under stress. In other words, at the emergency scene, you are more likely to forget what tasks you haven't done yet rather than what you've already completed. Because the tasks to be performed vary based on the type of emergency, it can be beneficial to have different checklists based on the various emergency situations you may face. This means the checklists for a vehicle accident, a structural fire, and a hostage situation would look very different. In the next chapter we look at the challenges responders face as they try to manage data and information in high-stress environments.

Fig. 12–3. Using checklists and worksheets to help manage workload is a situational awareness best practice. An aide or scribe to help document critical information relieves the commander of the task of having to remember to write information down. While documentation may appear to be a task that requires minimal cognitive effort, it can be very challenging and demanding under stress.

CHAPTER 13

DATA AND INFORMATION MANAGEMENT

Many emergency scenes are dynamically changing environments in which data and information management can create significant challenges. All the clues and cues that form situational awareness can be classified as data or information. Thus, when a responder struggles to capture, process, comprehend, and/or recall data and information, there is the potential for a situational awareness consequence.

Information Overload

First responders can find it challenging to manage the volumes of data and information that come their way during an emergency. In a dynamically changing, complex environment, it is easy to suffer from information overload. Rarely during my research on situational awareness or during my subsequent discussions with first responders did responders complain of the negative effects to their situational awareness of having *too little* information at the emergency scene to make a good decision. Conversely, they often complain that they suffer from information overload. They do note there are times when they didn't have enough of the *right* information or that they had more information coming at them than they could possibly process and make sense of.

Audible information

Everything a responder sees and hears is a clue or cue—information to be processed, including radio traffic. This can prove especially challenging for a command officer who must monitor radio traffic (figure 13–1). Unless there's someone listening to the radio on the commander's behalf, he or she must be attentive to every verbal radio message, and every one of those messages must be processed and comprehended. This processing and comprehending takes time and consumes precious, limited brain resources. When there are lots of radio transmissions, one of two things may happen: the responder may become so intent on listening to the radio traffic that important visual cues are missed, or the responder can be so intent on seeing the visual cues that important audible (radio) cues are missed. There's a third possible outcome: a responder can attempt to split attention between the two and, in the process, miss both audible and visual clues and cues (see chapter 9 for more on the myth of *multitasking*).

Fig. 13–1. It is a good situational awareness best practice for an incident commander to have someone else listen to the radio and then only share with the commander condensed versions of the most important pieces of information to aid in decision making. This helps the commander avoid information overload.

Visual processing of verbal messages

Your brain processes verbal messages as visual images, meaning that each verbal message becomes a mental picture. A first-arriving officer (or dispatcher) reports a certain condition or set of circumstances over the radio, and responders who hear the report form visual images of what they are hearing. This process takes time and consumes precious cognitive capacity. When the number of verbal messages coming in exceeds the brain's capacity to process the messages into images, some of the verbal messages may not be processed and comprehended into meaning.

Filtering out the noise

One of the challenges responders face with audible message processing is that the brain can't determine whether the messages are important until the sound waves are taken in and processed. The more messages that come in, the harder the brain has to work to convert those audible messages into something meaningful. When the brain reaches its capacity to process audible messages, additional messages may be missed. This is not only true for radio messages, but for face-to-face messages, too.

The formation of visual images from audible messages requires the brain to use imagination. Whereas the accuracy of the visual image can be improved by the quality of the audible description, radio transmissions are often abbreviated reports and paint only a partial representation of the facts. The commander or company officer can improve situational awareness and confirm their visual representations of audible messages by conducting a thorough size-up.

The size-up is one of the most important ways responders can capture (or confirm) important clues and cues. This involves a visual assessment of all sides of the emergency scene (when physically possible). The size-up is essential to the formation of accurate Level 1 situational awareness (i.e., capturing the clues and cues in the environment). When responders fail to complete a size-up, they are forced to rely on their imagination, extrapolated from verbal messages. This can contribute to a flawed Level 1 situational awareness. It also may, in turn, lead to flaws in Level 2 and Level 3 situational awareness.

No Size-Up

In general, many responders understand and appreciate the value of completing a comprehensive initial incident scene size-up. However, despite their acknowledgment of its importance, some responders fully admit they fail to complete this essential task. This may be due, in part, to the multitude of tasks they perceive need to be done immediately upon their arrival at an emergency scene. Sometimes they were in the process of conducting their size-up and got interrupted or distracted with something else happening at the emergency and never got back to completing the size-up. In some instances the building was so large that the responder could not physically maneuver all the way around it. Some noted that when incidents occur at night they might not be able to see everything they need to because of darkness.

Regardless of the reason, an incomplete size-up can prohibit a responder from gathering critical clues and cues, and this can flaw situational awareness (figure 13–2). Stated another way, if you don't collect useful and meaningful information, you can't figure out what's going on.

In addition to flawed size-ups, there are other ways commanders can fail to capture clues and cues. Responders and those who lead them often depend on the transfer of information via voice communications (either face-to-face or via radio). If another responder does not report the right information, accurately and timely, other responders may have their situational awareness impacted. You may recall from chapter 7, where I discussed communication barriers, that this can be a significant challenge for responders.

Fig. 13–2. A thorough and complete size-up helps paint a visual image for all responders of what they can expect upon their arrival. This serves two purposes: First, it helps responders mentally prepare for the incident. Second, it helps responders see what has changed from the initial size-up to when they arrive. This can help them determine the speed of the incident, an essential component of developing situational awareness.

The Role of Accountability

One critical piece of information required for the development of situational awareness and incident scene safety is crew accountability. It's often taught—and perhaps for the most part is commonly understood—that crew accountability is vitally important because if something goes wrong, fellow responders will know where to send rapid intervention teams to rescue the endangered responders. That's a given.

However, when it comes to the development and maintenance of situational awareness, crew accountability takes on a whole new level of importance. You may recall that the highest level of situational awareness (Level 3) involves being able to predict future events. You may further recall that one of the components of recognition-primed decision making is being able to construct mental models of future events.

So, to develop and maintain situational awareness and make quality decisions at emergency scenes, a responder must run mental models of future events. I liken mental models to little movies you play in your mind that help you look beyond the current events and make predictions about future events. As you're playing movies in your brain of future events, the *actors* in your movies are the first responders performing tasks at the emergency scene. And this is where accountability becomes catastrophically important to situational awareness.

To develop Level 3 situational awareness through mental modeling, responders must have information about the actors in the movie. Are they talented actors? Are they novices or expert actors? Are there enough actors to get the job done? And so on. A supervisor must possess information about the responders (the actors) who make the future events unfold with their actions (or inaction).

So what does a supervisor need to know about the actors? To answer that question, we again refer back to the conversation in chapter 6 where I discuss staffing as a barrier to situational awareness. The quantity and quality of the personnel operating at the emergency scene tells the supervisor not only *what* will happen but also *how long* it will take to happen. To make realistic predictions of future events, the supervisor must pay attention to how many personnel are assigned to perform a task, and the supervisor must make predictions about the ability of the crew to get the assigned task(s) completed in a certain amount of time. Why? Because that's what predicting future events is all about—to get a certain task done, to lead to a predictable outcome, in a set amount of time.

The completion of the task at a scene is how the proverbial movie plays out for real —not in your head. If the task gets done when and how it is supposed to, the movie plays out the way the responder envisions it would. If the actors don't perform the assigned tasks the way it was perceived or within the time anticipated, the supervisor's Level 3 situational awareness is at risk of being flawed (figure 13–3).

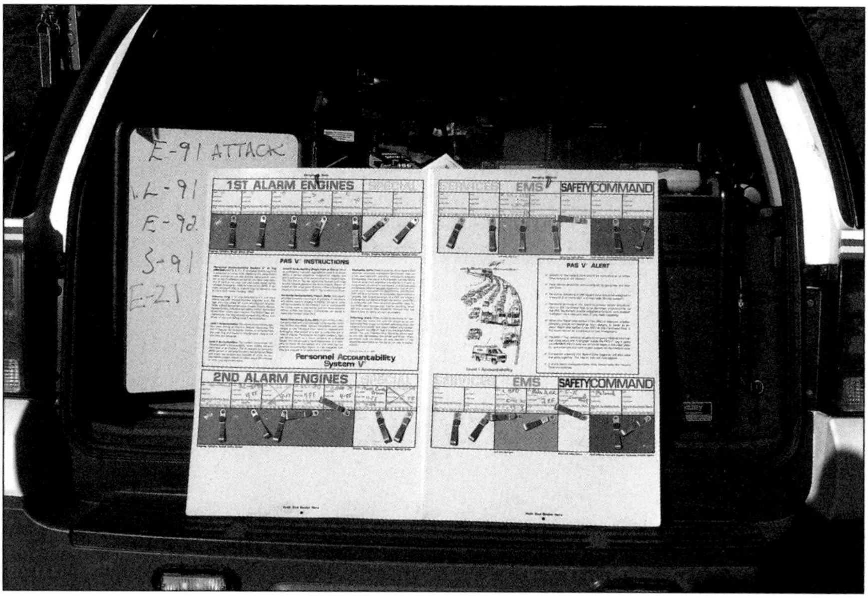

Fig. 13–3. In addition to the safety reasons for maintaining accountability, it also helps the commander maintain Level 3 situational awareness by knowing how each team's actions are working together to achieve the common objective. Knowing the crew size, location, and task helps predict future outcomes.

Size Up the Crews in Advance

One way for supervisors to improve their situational awareness is to improve their understanding of the abilities (and inabilities) of their personnel. Ideally, this would be done in advance of an emergency response. This can be accomplished by observing personnel while they are training and learning. In fact, supervisors who don't directly observe the performance and abilities of their crews are prone to having a flawed Level 3 situational awareness.

Let's face it (at the risk of sounding mean-spirited): Not all crews are equal in their abilities. The A-Team crew is going to be able to get much more done, more efficiently and effectively than the crew whose members look and act like they just arrived on the scene from the

Island of Misfit Toys. The supervisor must know something about the abilities (and inabilities) of each team member. This knowledge can be acquired by observing responders (the actors) during the rehearsal of their parts (during training). Being unfamiliar with the crews, the tasks being performed, and how long it will take to accomplish the tasks can seriously compromise the Level 3 situational awareness. To be able to predict the future, supervisors must intimately know their crew: size, location, task, and quality.

CHAPTER 14

MISSION AND GOALS

It's vitally important that all responders working at an emergency incident scene be on the same proverbial page. If personnel are unclear of the mission and goals, or if they are unfocused (and lose sight of the mission and goals), situational awareness can suffer. Incident scene safety is improved when everyone has a clear understanding of the mission and the goals to be accomplished.

Ideally, supervisors would articulate the mission, commonly called the *strategy*, to responders and set goals, which may also be called *benchmarks*. The mere act of getting everyone on the same page may be much easier said than done. (See chapter 7 where the challenges of miscommunications are discussed.)

If getting and keeping everyone on the same page was not challenging enough, mission and goal barriers can arise from the complications of rapidly changing conditions. As conditions change, supervisors must be resilient and nimble enough to adjust the mission and goals on short notice, oftentimes with incomplete information. They need to get everyone operating at the scene to understand the newly minted mission and goals while operating in a high-stress, complicated, often confusing and noisy environment.

I bow in respect to those supervisors who, when they see things getting worse, change the incident mission and goals to ensure the safety of their personnel. All too often this does not happen. Instead, supervisors can become locked onto their originally set mission and goals and refuse to adjust. This holds the potential for tragic consequence.

There are several reasons why this can occur. One explanation is the supervisor perhaps did not see (or comprehend) the conditions were becoming worse. As noted in previously discussions, paying attention is not as easy as it sounds.

Ego, Self-Esteem, Fear, Embarrassment, and Control

Unfortunately, sometimes a more sinister psychological trait may be the culprit—ego. A responder with a large ego (coupled with fragile self-esteem) may not adjust the initial mission and goals out of fear: fear of being accused of implementing a flawed plan, fear of failure, or fear of embarrassment for being perceived a failure. This ugly truth came out in my research. I interviewed supervisors who revealed that they knew their plan was not working, yet they stayed true to the flawed plan and the outcome was a near-miss or casualty event.

Assuming a supervisor is able to capture the clues and cues of the changing conditions, comprehend those clues and cues into meaning (understanding things are getting worse), and predict a future that includes a bad outcome, why would the supervisor remain committed to a flawed plan? The answer may lay in our understanding of a very powerful psychological drive—*control*.

An overly developed ego coupled with fragile self-esteem can be a recipe for disaster. As responders feel a need to control an incident scene, they may implement a mission and goals that are acceptable early on, but become stale and outdated as the incident deteriorates. As things change for the worse (i.e., the plan is not working), supervisors may lose control of the plan.

A supervisor may also get fixated on a single goal and lose sight of the fact that there may be multiple—and often competing—goals that need to be addressed simultaneously. A high-stress, high-consequence goal may narrow the attention of the supervisor (see the discussion in chapter 1 on attention barriers and the role of tunnel vision and task fixation). If a responder becomes task fixated, other important goals may be overlooked. For example, if a crew arrived at a working fire where a victim needs to be rescued, the supervisor might assign a crew to perform the rescue and then become fixated on those working the rescue operation. The responders may also fixate on the mission of rescue and ignore other important tasks. All the while the fire may continue to burn unabated, and the supervisor may forget to monitor the fire progression because it's not the goal.

To prevent goal fixation, responders should remain mindful of multiple, competing goals and strive to balance, or share, attention among several goals. Recall from chapter 9 where we discuss workload management barriers, that it's not possible to multitask and pay attention in a stimulus-rich, dynamically changing, high-stress environment. Thus, the attention to multiple goals is in sequential order, not performing tasks simultaneously. Paying attention to the status of multiple goals can be especially helpful as goal priorities shift. For example, if a crew is engaged in the rescue of a civilian victim while the fire grows to a point where crew safety becomes jeopardized from the fire, the priority goal may shift from rescue to fire suppression. However, if responders are fixated on the goal of rescue, important clues and cues about the growth of the fire may get overlooked.

> Ego may lead a supervisor to stubbornly think he or she developed a superb plan and will see that plan through to the end, regardless of how the developments in incident conditions dictate the need for a change. Think of it as Kamikaze incident leadership, except the supervisor may not be the direct casualty.

Goal Progress

Responders need to continually assess the progress of goals. In the complex and often chaotic environment of an emergency scene, a supervisor may set a goal (or multiple goals) and then simply fail to ensure appropriate progress is being made (or verify the goal was completed). If a supervisor sets a goal that a certain task is going to be accomplished in a certain period of time, it's imperative that he or she conducts a check when the allotted time has passed. If the anticipated progress has been made, it serves as a validation the supervisor's plan is working.

A warning flag should be raised if the expected progress has not been made and the supervisor should seek an explanation as to why the goal was not accomplished in the anticipated timeframe. This may be one of the greatest indicators of a flawed Level 3 situational awareness (being able to predict future events through modeling).

Part of the modeling of future events involves setting mental goals with anticipated deadlines (figure 14–1). If the goal progress is not aligning with the deadlines, something is wrong.

Fig. 14–1. The incident commander should develop and communicate incident goals and then track the progress of the incident to the goals. Benchmarking progress to goals helps commanders maintain situational awareness.

Standards

Another way responder situational awareness can be impacted is through challenges involving standard operating procedures (SOPs) or standard operating guidelines (SOGs). Whether you call them SOPs or SOGs (we refer to them as *standards*), these documents serve as the game plan by which a department operates under normal conditions. Not unlike one written for a football team, SOPs are the agency's playbook. The standards ensure everyone operating at the emergency scene shares a common understanding of how the emergency is handled under normal conditions.

> An organization that lacks written standards must rely on the development of common understanding of organizational operations by word of mouth. This can be a dangerous way to operate in a dynamically changing, high-consequence environment.

If responders are given the latitude to inject their *opinions* or *personal preferences* into how routine work tasks should be accomplished, supervisors may expect to see independent action by responders (i.e., freelancing). Further, supervisors should not be surprised when they observe wide variation in how incidents are managed. This variation, when combined with uncoordinated efforts, can increase errors, near misses, and casualty events.

Another barrier to responder situational awareness can occur when organizational standards are present, but personnel operating at the emergency scene do not follow them. Recall that responders build Level 3 situational awareness through modeling (predicting) future events (see chapter 4, "Situational Awareness"). These models are built, in part, on the responder's preprogrammed understanding of existing standards (figure 14–2). This sets up mental assumptions and expectations for how future events should unfold (which may include assumptions about the tasks certain persons or certain crews should perform). These assumptions and expectations, however, are not exclusively developed and held by supervisors. Everyone operating at the emergency scene forms assumptions and expectations of individuals, teams, and the entire incident. These assumptions are based, in part, on established standards.

Fig. 14–2. Standard operating guidelines establish expectations for job performance and duties in advance of an emergency. As responders arrive, they know what they should be doing under normal operating conditions. This helps in the development of team and incident situational awareness by creating a shared mental model (i.e., getting everyone on the same operational page).

Resilient problem solving

There are times, for sure, when there is a good reason for responders to *not* follow written standards. In fact, I strongly advocate for training personnel to understand how to be resilient problem solvers when faced with unique or novel issues that are not covered by written standards. When this occurs, it's imperative for the responders who are deviating from the standards to communicate to their supervisors their deviation in their course of action (away from the standards) along with an explanation for the variation.

> When something atypical is occurring that requires a unique or novel solution, communicating what that *something* is ensures that all the responders can adjust their expectations accordingly.

This communication is vital. It puts all responders on notice that things are not going to be routine (i.e., by the book). Absent this communication, arriving personnel (including the incident commander) may become confused as to why personnel are not following the script (the standard). If the situation being encountered is not covered by written standards, or if the problem is unique or novel and requires deviation from standards, the rule is *communicate!* This simple rule improves everyone's situational awareness.

Error creep

The final barrier addressed in this chapter is *error creep*. This barrier has been called by several different names by different researchers and authors, including:

- standardization of deviant behavior
- mission creep
- drifting into failure

Nomenclature aside, this is one of the most dangerous barriers to situational awareness because of the stealthy way it can implant itself within an entire organization.

Error creep occurs when an organization makes operational or safety mistakes, often small and seemingly harmless, repetitively over a period of time without consequence (figure 14–3). The lack of consequence from flawed operations or safety practices then leads personnel to a false sense of confidence that they are doing their jobs properly and safely.

The more times personnel are able to perform a flawed task without consequence, the more they believe the way they're doing the task is safe and acceptable!

Fig. 14–3. Repetitive performance of tasks causes responders to create habits that become automatic under stress. Building habits of performing tasks, such as ventilation while following best practices (e.g., wearing full turnout gear and SCBA), can help avoid error creep.

This confidence may further lead to additional error creep toward disaster as personnel continue to find additional ways to shortcut best practices, often in the name of expediency. Another consequence of error creep can be complacency—short-cutting standard procedures or best practices under the belief that no harm will come as a result. The more times a flawed procedure is performed without consequence, the greater the confidence becomes in performing the shortcut.

Then a casualty event occurs. The mere occurrence of the casualty may be enough to shock the organization into realizing they were suffering from error creep—but don't count on it. Sometimes when organizations have a high level of confidence in their flawed Standards they may fail to see the errors of their ways even in the throes of a casualty event. This may lead to an organization continuing to do things in the same flawed ways even after they have suffered a casualty event. They may be able to justify, in their minds, the casualty event occurred for some other, possibly unrelated reason, not because their organization is a victim of error creep. They drifted toward failure so slowly that they couldn't see it coming and then wallow in denial after it happens. As tragic as that sounds, I've witnessed it many times.

If the error creep results in a casualty, and the organization is in a state that falls under the jurisdiction of federal Occupational Safety and Health Administration (OSHA) regulations, the wake-up call may come in the form of a comprehensive OSHA review of departmental policies, procedures, and practices. That review may reveal the errors. Of course, this occurs only if the error is truly a violation of the OSHA law. It's important to acknowledge OSHA's charge is not to uncover or address operational flaws that are covered by state or local regulations.

A casualty event may also result in private civil litigation seeking culpability from organizational leaders who *allowed* the errors to creep into the organization in the first place. I add emphasis to the word *allowed* because it's possible that everyone in an organization may be proverbially blind to the error creep. If the civil litigation goes to trial, the command staff will have an opportunity to convince the court they were blind to their shortcomings. This may be difficult to prove.

Conducting Independent Safety Evaluations

If it's plausible that *everyone* in an organization can be blind to such errors, how can a leader remove the blinders and see the errors in time to prevent a catastrophe? The solution is rooted in a situational awareness best practice that you might use before you ever have an emergency: to perform a safety evaluation. I regularly urge leaders in emergency services organizations to do this, yet see far too few take the advice. To identify and correct error creep, an organization can agree to have a comprehensive, *independent third-party* evaluation completed. This independent evaluator should, ideally, possess considerable knowledge about best practices and be thoroughly versed on the contributing factors that lead to near misses, catastrophe events, flawed in situational awareness, and poor decision making. At the risk of sounding self-promoting, I would urge you to use this book as a resource to help you conduct such an evaluation of your organization.

The type of independent evaluation I am speaking of is different from the investigations an organization may be subjected to from OSHA or the National Institute of Occupational Safety and Health (NIOSH) on the heels of a catastrophic event.

For one thing, if you choose the right independent evaluator, the person or organization will serve as a resource for you—a friendly face, if you will—with the objective of trying to help you identify your errors and make recommendations for how to fix the errors *before* you have a casualty.

An investigation conducted after an event is not likely to be performed by friendly people with an interest in helping your department improve (figure 14–4). After a loss event the aggrieved will line up to prove your culpability and maybe even compensation for the loss. If your organization has unknowingly been drifting toward disaster (maybe even with an occasional near-miss event along the way), how would you want your errors identified and corrected? Would you want to learn from a friendly, independent evaluation process, full of great advice and best practices? Or would you want to have your mistakes pointed out by federal investigators with a citation book in hand and lawyers in search of responsible parties?

Fig. 14–4. Following a near-miss or casualty incident, the fire department may be involved in multiple inquiries to uncover lessons learned. Some of the investigators may be friendly and helpful. Others may be looking to place blame and find fault in the operations.

CHAPTER 15

MENTAL MODELS

Mental models are your mind's point of reference of what is happening now and what's going to happen in the future. At an emergency scene every responder forms mental models of what's happening. These mental models are based on a host of factors including the responder's collective training and experience, knowledge of the type of emergency (including building information), dispatch information, and size-up information. All of these factors combined can help a responder understand what is happening now and, equally important, what is going to happen next.

Mental models are rooted in situational awareness. Recall that Level 1 situational awareness is capturing the clues and cues happening in your current environment. Level 2 situational awareness is processing those clues and cues into meaning, and Level 3 situational awareness is making accurate predictions of future events.

In chapter 13 we discussed how mental models are formed and used to make predictions. Mental models are, in turn, a critical component of recognition-primed decision making (RPDM). If your brain forms a pattern match based on past experiences, it may trigger an intuition-based response. If that happens, you may get an intuitive gut feeling about what to do. You may not be able to explain how you possess this knowledge, because the gut feeling comes from your tacit (subconscious) knowledge.

No Intuition

There's a word of caution due here. The intuition formed from tacit knowledge is not flawless. The possibility exists that your brain may not make a pattern match (and thus there is no intuitive gut feeling about what you should do). Conversely, and perhaps even more insidious, your intuition—gut feeling—can be wrong.

You may recall from the discussion in chapter 13 that it is not necessary for a responder to have experienced the exact scenario being faced to benefit from a pattern match and the resulting intuition. How can you benefit from intuition if you've never had the experience before? Your brain can take bits and pieces of various past experiences (each of them stored in your brain as patterns), none of which, unto themselves, are an exact match to what is happening currently, and put the pieces of the proverbial mental puzzle together, forming a pattern match. This, in turn, results in an intuitive gut feeling of the appropriate action to take. Intuition (sometimes referred to in neuroscience as *unconscious decision making*) is a powerful decision-making tool for an emergency responder.

Intuition can, conversely, be a deadly nemesis. The brain, as magnificent as it is in processing information, is not perfect. You can get an intuitive gut feeling, based on a pattern match formed from past experience, guiding you to the wrong decision. Your intuition can be flawed. The brain forms pattern matches based on the clues and cues it processes. If there are important clues and cues missing, it's the equivalent to missing pieces of a puzzle. If your brain is processing the wrong meaning for certain clues or cues, it may get confused.

The brain cannot comprehend what it does not know. I have to pause here because that last statement is only *partially* true. The brain does have an amazing ability to take bits and pieces of information and, with remarkable accuracy, fill in the information that's missing. Indulge me for a moment as I digress with a brief explanation about the processing of visual cues (i.e., your eyesight). Understanding how your eyes work serves as a wonderful illustration of your brain's amazing ability to, well, make things up. Each of your eyes has a blind spot. Physiologically speaking, there's a small area in the back of the eye where there are no photon (light) receptor nerve cells. Thus, when you look at something, there's always a hole in your vision—a blind spot, if you will. How can this be? If that were true, logic would dictate that when you looked at something there would be two black holes representing the blind spots in each eye, right? This is where it gets amazing—and maybe even a little weird. Your brain fills in the blind spot based on your memory of past experiences, coupled with calculations it makes about your current surroundings. That's right. Your brain imagines what *should* be in your field of vision and fills in the blind spot.

This means, in essence, that you are actively hallucinating every day of your life. Don't believe me? See "the hallucination test" on the next page.

These hallucinations can be quite troublesome, not only at an emergency, but in daily life as well. If you're driving your car and another car is in your visual blind spot (not to be confused with the mirror blind spot), your brain may fill in your visual blind spot with the image of open road or an empty intersection. This can result in an accident or a near-miss event resulting in the proclamation, "*I didn't even see the other car*." Maybe you simply weren't paying attention, or maybe you truly didn't see it because the other vehicle was in your *blind spot*.

Similar to the way the brain resolves the blind spot issue, when faced with incomplete information, your brain can fill in the facts based on what it thinks the facts should be, giving you an intuitive gut feeling based, in part, on incomplete inputs (clues and cues). The odds are in your favor that your brain will be accurate (or close enough) to trigger an accurate response based on intuition. But what if it's wrong? What if your cognitive blind spot was filled in with a proverbial blank roadway when there was actually a tractor and trailer coming through? The results can be disastrous.

The whole idea of faulty mental models, blind spots, hallucinations, flawed pattern matching, and defective intuition may be enough to scare you into considering changing vocations. But, before you take such a drastic step, know there are some things you can do, in spite of all these obstacles, to improve your situational awareness and, subsequently, your decision making.

The Hallucination Test

Take a blank index card and on the left side, put a dot the size of a pea. On the right side, put an X about the same size (see the example in figure 15–1). Hold the card in your outstretched left hand so the X is on the right side of the card. Keep both eyes open, but use your right hand and completely cover up your right eye. Look at the X using your left eye. While your eyes are focused on the X you should still be able to see the dot in your peripheral vision. Concentrate on the X while slowing drawing the card toward your face. As you draw the card toward your eye, at some point, the dot is going to disappear. When it does, you have discovered your *blind spot*. If it doesn't happen, try it again. If you can't make it happen at all, then you may be one of those rare, anatomical freaks of nature, and you may want to consider signing up for the circus.

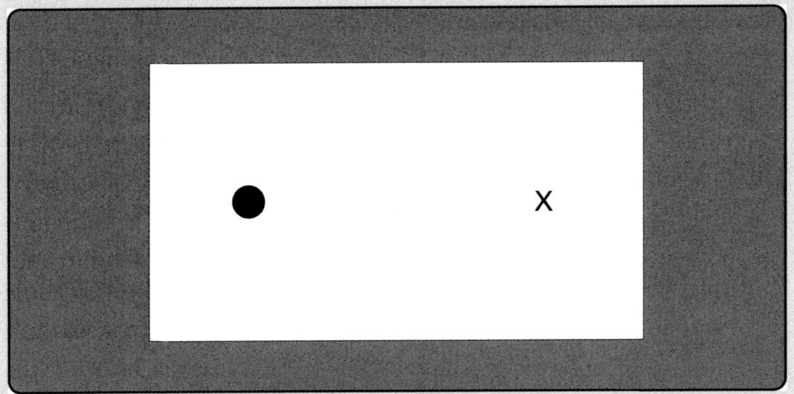

Fig. 15–1. The hallucination test

Do this exercise and you will quickly see you have a blind spot and understand your brain's ability to fill in missing information—to hallucinate.

You may recall from our discussion in chapter 3 that one of the four essential requirements for making a good decision is situational awareness. In turn, to develop and maintain strong situational awareness requires three things of the decision maker: You must be conscious (or awake), you must be present (which can include being physically absent from the scene yet present where radio traffic is providing critical clues and cues), and you must scan your environment (both visually and audibly). Scanning the environment serves two important purposes. It helps responders avoid fixation on any one task or geographical area of the scene, and (germane to this discussion) it helps fill in the blind spots with accurate information.

Trust, but Verify

Another critically important action responders should take that will improve the pattern matching ability of the brain is to search out *facts* that validate the accuracy of intuition. Facts are the evidence a rational brain can use to validate the veracity of intuition.

To blindly trust intuition with no form of validation can be foolish, if not fatal. Borrowing the words from former President Reagan in the context of his dealing with the Russians on the development of a nuclear arms reduction treaty, "*Trust, but verify*." This is a good rule to apply when managing your intuitive gut feelings under stress.

Your brain can be a stubborn brain

As so often is the case throughout this book, I have to add another unpleasant caveat to your efforts to validate intuition with facts. After you lock onto a belief, it can be difficult to change your mind. The brain can be frustratingly stubborn in this regard, even in the presence of facts that run contrary to the beliefs of the responder. There is a risk that those facts may get dismissed because they don't fit the responder's understanding of what is happening. The responder's brain may make up some excuse as to why the new information cannot possibly be accurate, or may dismiss the facts as inconsequential. Thus, a responder may overlook critical, factual information if his or her mind is already set on what it believes the situation *should* be.

We see what we want to see

If it weren't bad enough that your brain can dismiss facts that don't fit your understanding of the situation, it can get even more complicated. Armed with a flawed understanding of what is happening, a responder goes on a fact-seeking mission. However, the brain can unknowingly deploy a vetting process that seeks out facts that only *support* the existing beliefs. In other words, a responder may not even see or hear cues and clues (i.e., *facts*) that don't *fit* the mental model. Perhaps stated another way in a lesson that can be applied on an emergency scene as well as in our daily lives—*we see what we want to see*.

Challenge and verify

You can help avoid this shortcoming by playing the role of the devil's advocate to your intuition. Instead of dismissing errant facts out of hand, ask yourself how those odd clues and cues may mean something different than what you think is going on. Ideally, I would advocate for an entirely separate person to serve this role, someone to challenge and verify the veracity of your mental model. This is where the talent of an experienced command aide or advisor can be crucial (figure 15–2).

The challenge and verification process (challenge my decision and I'll verify what I'm seeing with evidence) can be valuable in the process of making decisions under stress. But it must be done in moderation. Ideally, however, you would do this before your incident starts running off course and going in an unsafe direction based on a flawed mental model. Oftentimes there simply is not enough time (or enough mental capacity) at an emergency scene to run a laundry list of what-if scenarios for extended periods. The badgering of a decision maker in this fashion may serve only to choke his or her ability to make and implement a timely action plan.

Fig. 15–2. A command aide can be a tremendous asset to an incident commander. Aides provide a second set of eyes and ears to capture and process clues and cues. Their findings can then be compared to the commander's. Discrepancies should warrant further evaluation.

The pre-arrival mental model

It's not uncommon for responders to form a mental model of an emergency scene prior to actually arriving. This can be done by combining information about the current call with prior experiences of similar types of calls or previous calls to that particular address.

When I was a young responder I was taught that when I respond to a call, prepare myself mentally for the worst possible scenario I could find on arrival. The logic behind the advice was mental preparation would set me up to be ready if the actual scene was really *that* bad, and it would calm me if I arrived and the scene was not as bad as I had envisioned. I always assumed the worst and hoped for the best. That advice proved to be valuable for me throughout my years as a responder and commander.

Applying this advice resulted in my being well prepared for any challenge the incident scene held for me.

However, the formation of mental models in advance of arriving at an emergency can have a dangerous side effect. Responders can form a mental image of what the emergency scene is going to be like and make decisions about what they're going to do (or even give orders to other responders as to what they are to do) before ever arriving to see if the real incident aligns with the imagined incident (figure 15–3).

Fig. 15–3. The pre-arrival mental model, based on radio transmissions, must be compared with the actual clues and cues after arrival. Failing to do so will result in a flawed Level 1 situational awareness.

Living in a fantasy world

I have spoken to responders who have admitted that, while responding to an emergency, they formed mental models and used their radio to give other responding companies orders of what to do when they arrive.

Upon arrival at the scene, and for reasons the responders often struggled to explain, they admitted remaining true to the mental model (and the decisions) they formed prior to arrival, even if the actual emergency scene was different than they had imagined it to be. Why would someone do this? The imagination is very powerful, and the brain has a difficult time trying to distinguish facts from vivid imagination. If a responder vividly imagines something to be factual, the brain may process the fantasy as reality. When the responder arrives, the real clues and cues at the emergency don't fit the fantasized reality. After you make up your mind to do something (even if it's before you arrive, and even if it's based on flawed mental models), it can be difficult to change your mind.

CHAPTER 16

COMMAND LOCATION

The focus of this entire book is situational awareness, with a heavy emphasis on the barriers (i.e., impediments) to forming and maintaining situational awareness. Some of those barriers are interrelated. I will stop short of inferring that there is a cause and effect relationship among various barriers because that would be an assertion I have no scientific backing to support. However, responder participants in my research noted, without hesitation, that while being affected by one barrier, another barrier was simultaneously causing them problems.

For example, some responders noted their situational awareness can be affected when short staffing causes them to assign high-risk tasks to understaffed companies. Under these conditions, responders note they tend to focus their attention on the high-risk tasks being performed by the understaffed companies (leading to task fixation or tunnel vision). Simultaneously, some of these responders noted the high-risk short-staffing situation caused them to feel anxious, angry, and/or frustrated, and each of these emotions is, unto itself, a barrier to situational awareness.

Location, Location, Location

The commander's location is another example of a situational awareness barrier that can be simultaneously influenced by a variety of other barriers. The heart of the issue here, holding no punches back, is the debate of where an incident commander should be physically situated during an emergency. For the sake of this discussion I limit the options to two choices: inside a vehicle (what I call being situated in the *seat*), or outside of a vehicle (what I call being situated in the *street*).

Commanders can be passionate about where they think they should be located. Street commanders justify their position by noting they must be out in the action where they can see and hear everything. If they need to talk to a responder, supervisor, homeowner, another public safety provider, or utility company worker, they can do so face to face. If they can't see something that's happening, they can physically move to another vantage point where they can directly assess what is going on (figure 16–1). They said things like, "I can't be caged up inside a vehicle."

Fig. 16–1. Some commanders prefer to position themselves close to the action where they can have immediate face-to-face interactions with personnel on the scene.

Conversely, seat commanders note they prefer to be removed from the action to avoid seeing and hearing everything. (This is just about the exact opposite argument made by the street commanders.) Seat commanders note they want to avoid having their situational awareness impacted by interruptions from face-to-face discussions with responders, supervisors, homeowners, other public safety providers, or utility company workers (figure 16–2). They said things like, "I have to be in a quiet environment where I can calmly process information and keep track of what's happening."

Fig. 16–2. Some commanders operate inside a vehicle, preferring to be isolated from the action.

So who's right and who's wrong? That's a hotly debated question and one that begs for scientific research to determine the correct answer. Absent the science to validate the efficacy of command location, all I can do is share the observations of the commanders I interviewed and observed, and infer from those experiences where command should be situated.

Commanding in the streets

Some street commanders acknowledged their situational awareness can be adversely impacted from being situated in close physical proximity to the action. Their location exposes them to distractions like loud noises, bright lights, moving objects (e.g., vehicles, hoses, and people), and interruptions from persons approaching them for face-to-face communications. Street commanders also noted they can be prone to missing radio communications while operating in a loud, high-stress environment. Street commanders also acknowledged it can be difficult to use command worksheets and checklists, depending on weather and lighting conditions.

Nonetheless, despite all of these potential barriers, some street commanders acknowledge they rarely retreat to a vehicle, even after they admit knowing their situational awareness has been impacted by barriers related to their location.

Some street commanders articulated the application of rules of thumb they apply to help them decide when to physically move back from the action or retreat to a vehicle. Those rules may include when there is a known victim rescue or when the incident goes to a second alarm. Although these rules existed in the minds of some street commanders, the same responders admitted there were numerous times when they knew they should have physically moved back or retreated to a vehicle . . . but alas, they didn't.

Street commanders are more susceptible to distractions from the weather as opposed to seat commanders. A commander who is hot and sweating or a commander who is ice cold and shivering isn't going to be able to concentrate as well as a commander who is operating in a comfortable physical environment. Rain or snow can impede a commander's vision and also impact the effectiveness of worksheets and checklists. Paper and computer-based aids don't hold up well in wet environments either. Laminated boards that require the use of felt markers also do not hold up well because the ink can blotch and run in rain or snow. Wind and rain conditions can also impact a commander's hearing (figure 16–3).

Fig. 16–3. First responder situational awareness can be affected by extreme weather conditions.

Commanding in the seat

Some commanders are comfortable performing their roles in the confines of a vehicle. This location can have some distinct benefits, including operating in an environment that protects the commanders from distractions and interruptions. This environment can improve monitoring of radio traffic and tracking of personnel using command boards or computer-based technology. Commanding in a vehicle, where the temperature and lighting can be controlled, can also be more comfortable.

There may be disadvantages to commanding from a vehicle as well. If the commander does not have someone to serve as front-line eyes and ears, critical information can be missed and situational awareness can erode. A strong argument could also be made that humans evolved to be exquisitely tuned in to signs of danger—a skill that was not acquired while sitting in the cave. Rather, primal perceptions evolved while our ancestors were, for all intents and purposes, in a constant state of movement. Arguably, moving around an incident scene might improve the ability to capture and process clues and cues (with consideration to how distractions and interruptions might offset these benefits).

Every Location Has Its Challenges

In fairness, some commanders who situate in the seat have acknowledged their situational awareness can be impacted if they arrive first and do not (for whatever reason) conduct a complete 360-degree size-up of the incident scene. Seat commanders also noted their vehicle location is not always ideal for observing fireground operations. When in this situation the commander may rely on verbal reports from subordinate officers. Unfortunately, those verbal reports may not be complete or accurate, which can impact the commander's situational awareness.

If a commander takes up a fixed position in the cab of an apparatus, the noise of the engine may create a significant audible distraction for the commander. Wearing a radio headset to reduce ambient noise may reduce this barrier. The seat in the cab of a vehicle may also become uncomfortable for a commander over an extended incident. Discomfort can, in turn, erode situational awareness.

As for my personal experience, I have commanded from both positions (street and seat). In my early years as a commander no one could have ever convinced me there was any other place to command from than in the *street*. I had to be in and among the action, which unfortunately, included plenty of opportunities for me to perform hands-on work while I was, allegedly, commanding. From my interviews and backed with my research in cognitive neuroscience, I now understand the temptation to become a *hands-on commander* was likely due, in part, to my hereditary stress reaction that drew me toward doing the tasks I was most comfortable doing. For me, at that time (as a newer commander), my comfort was definitely being in and among the action and hands-on. I had not yet developed a comfort with being hands-off. When I wasn't hands-on, I felt guilty that I wasn't helping my crews with the hard work of fighting the fire. It was easy for me to invent every possible excuse and explanation for why I needed to be hands-on.

However, as time passed and the number of incidents I commanded increased, I gained experience and confidence in the act of commanding. This allowed me to see how being close to the action almost automatically meant more distractions, more interruptions, and a greater negative impact on my ability to do the job I was *supposed* to be doing (commanding the entire incident from a big-picture perspective). The more I realized I needed to be physically removed from the action, the easier it became to justify in my mind it was the right thing to do for the safety of the firefighters operating under my command. Unfortunately, that valuable lesson has been lost on many commanders who believe they must be, for whatever reason they want to invent, hands-on.

I can honestly say the transition from being a street commander to being a seat commander was challenging for me. Early on, even when I knew I should be in a fixed position inside a vehicle, I found myself getting out of my vehicle to address problems and issues with personnel. Sometimes I made it back to the vehicle, and many times I did not.

Eventually, after lots of practice and a great deal of self-discipline, I was able to keep myself in the seat and developed the habit I believe reduced my distractions and interruptions and vastly improved my situational awareness and decision making. If I was first to arrive, I conducted my 360-degree size-up before assuming my fixed, remotely located command position. If a company arrived first, I depended on the company officer to conduct the size-up and accurately report his or her findings to me.

Surrogate eyes and ears

I often found it beneficial at larger and more complex incidents to have an operations officer and a safety officer to serve as my forward set of eyes and ears, capturing the clues and cues up close and personal, processing the clues and cues into meaning, and communicating back to me the critical (few) pieces of information I needed to make good strategic-level decisions (figure 16–4). Operations and safety officers who are able to capture, synthesize, and communicate the essential information to command are incredibly valuable assets. Conversely, an operations or safety officer who does not possess this skill can adversely impact a commander's situational awareness by leaving out important details, relaying the wrong information, or sharing too much information.

Fig. 16–4. A safety officer or operations officer can be a valuable asset to a commander when in a position to capture, filter, and communicate essential information back to command.

The Laws of Work Interruption

Regardless of where a commander chooses to locate, four facts are indisputable and have been verified by scientific research. I call them the *Laws of Work Interruption*.

- **First Law of Work Interruption:** Interrupted workers make more mistakes.
- **Second Law of Work Interruption:** Interrupted workers miss more clues and cues.
- **Third Law of Work Interruption:** Interrupted workers take longer to complete their tasks.
- **Forth Law of Work Interruption:** When interrupted, a worker is likely to lose track of the task being performed prior to the interruption.

Clearly, when the commander's work is interrupted, the impact can be significant.

In discussions with street commanders, I inquired if they would ever command from a vehicle. Many reported to me that when an incident becomes complex (e.g., second alarm or a Mayday is called) they would physically back away from the action or get inside a vehicle where it was quiet. They reported that getting away from the action allowed them to control distractions and interruptions. The remote location gave them the breathing room to start writing things down, and they were better able to keep track of things (figure 16–5).

Fig. 16–5. A commander working in an isolated environment, such as a command vehicle, can control many of the distractions and interruptions that can hinder situational awareness.

The challenge for a street commander comes with knowing when it's the right time to move back and command from a position remote from the action. Many street commanders admitted to me that by the time they realized they needed to move back, they were already overwhelmed, and the incident's complexity was impacting their situational awareness. As one commander told me: "By the time I made my way to a vehicle, I never got caught up. I was never able to get the command board filled out. I'd already forgotten much of what I needed to write down. I couldn't remember where my companies were, what they were doing, or what their assignments were."

When things are going well and the complexity of the incident is low, it may be easy to keep track of everything in your head. However, as things become more complex and more personnel are assigned to varied tasks in multiple locations, the ability to keep track of everything can be significantly diminished. Unfortunately, when you start getting overwhelmed with incoming information, your brain can start shedding data (i.e., you start forgetting). And there is no assurance the things you do remember will be the most important.

When you get overwhelmed with information, you won't be able to take in and process the meaning of all the clues and cues you're being bombarded with. And when you get behind, it can be hard to catch up because the new information may not make sense without the knowledge of the previous information. It can become a vicious cycle, spiraling you into a situation where you have no awareness at all about the events that are happening. Catching up can be challenging.

There are two strategies that can help you manage the impact of being overwhelmed with information: writing notes about the situation and using checklists. By writing down the most important information, you ensure it won't be lost when you get overwhelmed. As you see or hear the critical clues and cues, capture them on paper. This helps immensely with managing your cognitive load. The more you write down, the less you have to remember. This is why some commanders use worksheets (figure 16–6) to help manage information. The worksheet helps improve retrospective memory—the memory of everything you've already seen and done. The second strategy is to use checklists. As you find yourself in a situation where the sheer volume of information coming in is overwhelming, it's easy for you to not only forget what's happened, but to also forget the tasks that haven't been done yet.

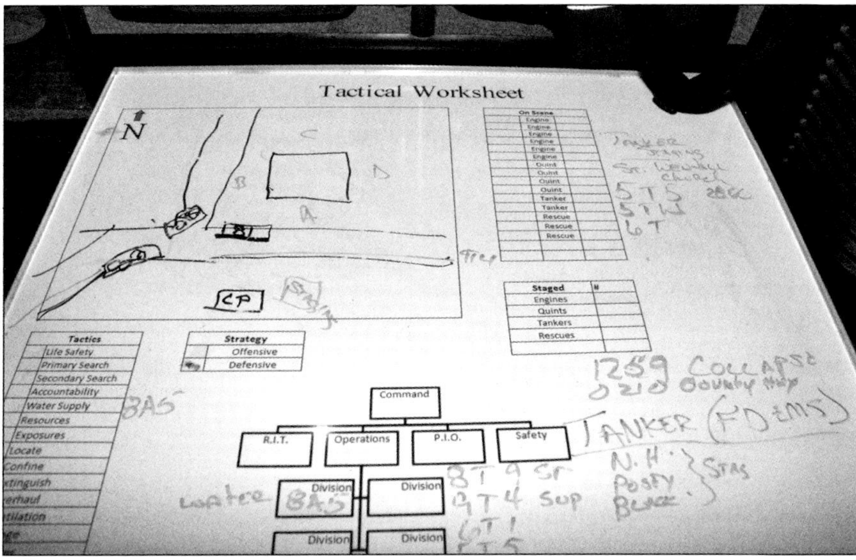

Fig. 16–6. A commander should use a tactical worksheet to document details of the incident. Keeping mental track of multiple company locations, crew size, and activities under stress is not realistic and can become especially challenging if an incident deteriorates rapidly.

Checklists help you manage prospective memory—the memory of everything you have not yet done. This is why people make grocery lists. The checklist serves to prompt the reader about what needs done, or at least considered. It's a cheat sheet, and if there were ever a place or time to say it was acceptable to cheat, it would be in using aids to help remember everything that needs to be done in a high-stress, high-consequence work environment (figure 16–7).

Another strategy to help manage information is to avoid allowing your attention to be drawn toward objects and events with which you should not be concerned. In other words, if there is someone else on the emergency scene whose job is to secure and maintain the water supply, do not expend your cognitive capacity on water supply issues. The more your attention is distracted and your work interrupted, the more vulnerable you are to overload.

Residential Dwelling Fire Checklist
(Not necessarily listed in priority order)

__Adequate resources dispatched	__360-degree size-up
__Establish water supply	__Victim survivability profile
__Safety	__Operations
__Search & rescue	__Attack
__Backup	__RIT
__Vent	__Entry
__Ladders	__Utilities
__Rehab	__Staging
__Electric company	__Gas company
__Victim welfare	

Situational Awareness Matters!

1. Capture clues and cues
2. Understand what is happening right now
3. Make accurate predictions about future events

7 Critical SA Components

1. Know where all my personnel are at all times
   - Crew size
   - Crew quality
   - Location
   - Task
   - Progress
2. Keep track of the passage of time and how fast conditions are changing
   - Smoke
   - Fire
   - Building deterioration
   - Wind
3. Control distractions and interruptions
   - Location
   - Environment
   - Noises
   - Bright things
   - Movement
   - People in my proximity
4. Keep a big-picture view of the incident
   - Capture clues and cues
   - Avoid task fixation
   - Keep out of the action
5. Set smart and achievable goals
   - Enough people
   - Enough resources
   - Enough water
   - Enough time
6. Ensure the benefit justifies the risk
   - Victim survivability profile
   - Firefighter survivability profile
7. Share the mental workload
   - Operations
   - Safety
   - Aide
   - Advisor
   - Worksheets
   - Dispatcher

Fig. 16–7. A commander can use a checklist to help manage prospective memory—the memory of everything that needs to be done at an incident. The checklist serves as a visual prompt, ensuring that the commander does not have to recall everything that needs done out of memory.

> One of the most insidious facts about situational awareness is you may never know you're losing it until it's too late. There's not going to be any trumpets blaring or neon lights flashing to signal your loss of situational awareness. In fact, in most instances, it's a gradual degradation over a period of time.

The length of time it takes for situational awareness to erode varies. It's possible for it to happen over just a few moments, and it's equally as possible to happen over days or weeks, which may be a concern for extended operations such as wildland incidents.

Meta-awareness

This conscious awareness of your situational awareness has a name. It's known as *meta-awareness*. You develop meta-awareness by figuratively stepping outside yourself and looking at your own level of awareness as if you were a third-party observer/evaluator. Rate yourself. Conscious (meta-) awareness is a hallmark of those who can maintain strong situational awareness. They use an internal assessment tool to rate their own performance in real time. This helps them see if their situational awareness is weak or strong.

> If it were possible to attach a Post-it note to your brain, the following might serve as a reminder for you about how to develop and maintain strong situational awareness:
>
> - Ensure you are capturing the most important clues and cues.
> - Understand the meaning of those clues and cues.
> - Make reasonable predictions of future outcomes.

There are probably few better ways to develop strong situational awareness than imagining yourself on the horizon before you actually arrive there. This is the essence of Level 3 situational awareness—the ability to predict your future events. The more you get distracted or interrupted, the less likely you are to have the cognitive capacity to make accurate future predictions. You're too busy getting back on task after being distracted or interrupted.

First responder casualty reports amply document incidents where there were all kinds of signs and indications (clues and cues) that the incident was heading in a bad direction. Being overloaded and/or overwhelmed impacts your ability to develop and maintain situational awareness, and you may be criticized (maybe even cited by investigators) for missing important clues and cues, or for failing to comprehend their meaning. It never happens on purpose, but nonetheless, it happens and it's avoidable.

A hundred paces

Another way to help manage distractions and interruptions is to remove yourself from the very environment that's causing you to be distracted and interrupted. You can accomplish this by simply taking a step back. Okay, more than "a step" back. I should say "take a hundred steps" back. In other words, get out of the action. Putting distance between yourself and the incident not only helps you physically see the *big picture*, it also removes you from the people, noises, bright lights, and activities that can draw your attention away and impact your situational awareness.

If you've ever watched a professional football game on television, one of the first things you notice about the head coach is that he wears a headset. Have you ever wondered who he's talking to? He's talking to a coordinator, usually located in a booth at the top of the stadium at the 50-yard line. The physical attributes of the booth are conducive to maintaining strong situational awareness. The booth is away from the action; it's also located where the coordinator can see the big picture. Coincidentally, the booth has well-controlled lighting and temperature to ensure the comfort of the coordinator.

Think about this for a moment—the head coach is on the sidelines. There are players all around the coach, many of them having conversations with the coach. The opposing team members are moving in close proximity, too. Thousands of screaming fans can overwhelm the coach's ability to hear (and perhaps think clearly). It can be cold outside—really cold! Ten degrees. And to make matters worse, it may be snowing, obscuring the coach's vision as melting snowflakes make water spots on the coach's eyeglasses. The coach's footing may be slippery under the accumulating snow, causing him to walk gingerly. The coach's hands may be cold—almost numb. The moisture in the coach's nose freezes with every breath taken. The only objects keeping the coach from getting frostbitten ears are the muffs on the radio headset. On the other end of that headset is the coordinator—sitting on a comfortable chair in a quiet, warm, well-lit work environment with access to playsheets and all sorts of data within ready reach and, most importantly, a big-picture view of the field and all the plays. It doesn't take a rocket scientist to see which of these two people are in the best position to develop and maintain strong situational awareness about the game. The coordinator's task is a formidable one—to keep the coach on the field informed enough about what is going on to help the coach make good decisions.

In an ideal world, it would be the head coach in the booth, and the underpaid, under-appreciated underling would be down on the field. That would make more sense. But, alas, professional football is first and foremost entertainment, and fans pay lots of money to attend professional football games. They expect to see the *actors* in the *show*, and the coach is one of those actors the fans want to see. If the coach remained in the booth, it would be a better position for seeing and hearing the essential things necessary to make good decisions. However, the fans wouldn't be very happy if they didn't see the coach functioning in his apparent high-stress work environment. The coach knows enough to have a coordinator up in the viewing booth to see the big picture and report essential information and observations through the headset.

There's a lesson in the football analogy for first responders. It's absolutely essential to have someone watching the big picture. Ideally, this person is located at a distance far enough away to be able to see the big picture. To be physically removed from the action facilitates the ability to see the big picture. Physical distance from the incident also affords a commander the ability to be clear about the things that cause distractions and interruptions. Avoiding distractions and interruptions is a critically essential skill for developing situational awareness and facilitating quality decision making.

CHAPTER 17

TEAMWORK

As if it weren't bad enough that responders can have flawed situational awareness and trouble making sense of what's going on around them because of individual challenges, the problem can be compounded when there is more than one responder looking at the same thing and trying to make sense of the situation (figure 17–1). It's possible that each of them may see something completely different. While this can be an asset at times, it can also be a tremendous liability.

As a refresher, making sense of something is the essence of situational awareness—capturing the clues and cues in the environment (Level 1 situational awareness) and comprehending those clues and cues into meaning something coherent (Level 2 situational awareness). But what happens if two responders look at the same situation and don't see the same thing? Maybe each is processing different clues and cues.

Fig. 17–1. It is entirely possible for two responders to look at the same thing at the same time and see entirely different things. When this happens, they are said to be lacking a "shared mental model." When this happens they should discuss what each other sees, not in a judgmental way, but in a way that seeks mutual understanding.

Negative Clues and Cues

Perhaps one responder's area of expertise lends itself to processing the meaning of *missing* clues and cues (known as *negative* clues and cues) whereas the other responder does not have the same level of expertise and, therefore, can't make sense of the same clues and cues. Recall that in the mind of a novice, clues and cues that are absent have no meaning. But to an expert, missing clues and cues can provide additional critical information about what is happening (or not happening). Regardless of the cause, when two responders do not share the same meaning of what's happening in their environment, it can be problematic in a big way (figure 17–2).

Fig. 17–2. Most novice responders are able to see and hear positive clues and cues. However, it takes expertise to comprehend "negative" clues and cues that are absent from an incident scene. Negative clues and cues can mean as much as, and sometimes more than, positive clues and cues.

On several occasions I've had the opportunity to run a situational awareness exercise that demonstrates this point nicely. I show a group a picture that can be interpreted multiple ways. Inevitably, some see one thing and others see something else. Where it gets really interesting is when I select two people from the group who see different things and ask them to describe to each other, in vivid detail, what they see. No matter how much detail one uses to describe the picture, it's often the case that the other can't see it. In fact, it's been my observation that the greater the detail of the description, the more difficulty the other person has with seeing the image their counterpart is describing. It seems counterintuitive, but it's exactly what happens. When a responder assesses a clue or cue (in this case, a feature in a photo) and makes a conscious decision as to what that clue or cue represents, it can be a challenge for them to see the clue or cue as anything else. Once formed, a perception can be difficult to change.

The exercise gets even more interesting, though. After letting these two banter back and forth for a few moments, and their frustration levels rise, I used a pointer to highlight the exact features that each had, up to that point, been described only verbally. You might think that after the essential features were pointed out, each would have an epiphany and see the alternative picture. Sometimes they do, but many times they don't. On occasion, one of the two participants will succumb to the stress and embarrassment and say, "Okay, now I see it." Then I'll hand over the pointer and ask one person to point out the features the other person described. That stops them right in their tracks. They can't do it. The point of the exercise is not to embarrass anyone. The point is to demonstrate how powerful (and sometimes inflexible) individual perceptions can be.

Again, after you make up your mind to see something a certain way, it can be very difficult for you to see it any other way. Let's apply this concept to an emergency scene.

Two People Can See Things Very Differently

A responder arrives at an emergency and develops his or her situational awareness by capturing clues and cues and processing them into meaning. The responder has, in essence, made sense of what has been seen and heard (and perhaps tasted, smelled, and felt, as well). Then a subsequent responder arrives at the scene and does the same thing, only this responder processes the clues and cues into meaning something different. Who's right and who's wrong?

Unfortunately, the answer to that question is often a function of the rank or official position in the organization's command structure. The problem is that people of low rank (or who lack the years of experience to have enough seniority, or who have not had opportunities to prove themselves and thus lack respect from peers and superiors) may never be given the opportunity to explain that they see something that appears to be very different from what their superiors see. The default of who is *right* may go to the responder with the highest formal rank. And even if the responder was given the opportunity, the outcome can be the same as the situational awareness exercise discussed earlier in this chapter. The supervisor may not see it, even after the second responder meticulously points it out.

Ego and self-esteem can also cause problems

A responder with good self-esteem and a healthy ego might consider this difference in sense making a significant warning sign that something may be wrong with his or her situational awareness (figure 17–3). Conversely, a responder with an exaggerated ego and low self-esteem may dismiss the observations of the second responder as inaccurate. As discussed in chapter 10, a balanced ego and self-esteem facilitate the ability to admit fallibility in sense making. Low self-esteem and an over-inflated ego can suppress the acknowledgment of fallibility and can lead to catastrophic outcomes.

Fig. 17–3. Great commanders (those with a healthy ego and strong self-esteem) will always welcome the input and good advice from aides and advisors. They know there are too many things they can miss, and the results can be catastrophic.

Uncoordinated sense making

I have interviewed command officers who admit to operating on emergency scenes where additional personnel arrived and the commander did not give the additional responders instructions on what to do. These commanders might say, "We have good SOPs, and the arriving personnel know what to do. I don't have to tell them what to do." Essentially, these commanders hold an expectation that each additional responder arriving at the scene will be able to conduct his or her own independent size-up of the situation, see all the same things as the previous crews (and the commander), and automatically know what to do in a coordinated fashion with other responders whose understanding may be very different.

It's challenging enough for two responders to arrive at the same time and have them agree on what they see. When you add the element of elapsed time between the arrival and size-up conducted by the different responders, the possibility of each having different mental models, and subsequently different situational awareness, can exponentially increase. Why? Dynamically changing incidents change as time passes, which means the clues and cues (the foundation of sense-making) are going to be different.

The commander doing the job properly should be watching and listening to the overall incident, seeing the progression of events and the changing clues and cues. This puts the commander in an infinitely better position to direct the activities of the additional arriving responders. Not only does the commander have the benefit of seeing the big picture, but the commander also has the ability to watch the conditions change and gain a sense for the overall speed by which the incident is progressing.

Say you are watching a sporting event on television, and a friend enters the room. Your friend has no idea what's happened in the game prior thus far. All the friend can do is surmise what's going on based on what can be seen and heard at that moment. If the event were, for example's sake, a football game, he or she might be able to determine which team has the ball and what yard line they are on, but not much more. The on-screen score board might reveal the score and what down it is, and how many yards are left to the next first down. But your friend is going to be unaware of some important information that occurred earlier in the game. Perhaps the opposing team's star quarterback got injured on the previous possession, but that team is on defense now and your friend is unaware that happened. Suffice it to say a lot has happened in the game prior to the friend's arrival that affects his or her sense making.

Teamwork and mutual understanding are critically important to first responder safety. Ideally, responders would capture and comprehend the same clues and cues the same way, and predictions of future events would be aligned in the minds of all responders and those who supervise them. This is a big task to accomplish in a dynamically changing environment, yet it is so important. Oftentimes the after-action reports from casualty incidents reveal that the responders did not have the same understanding of what was going on, what was being done, and what their role meant to the overall mission of the incident.

PART 3

LESSONS AND BEST PRACTICES

CHAPTER 18

LESSONS FOR FIRST RESPONDERS

Congratulations! If you've read the book to this point I am hopeful you have had a couple of situational-awareness, decision-making, and perhaps life-changing epiphanies. If you're feeling like I did when I uncovered all of these barriers to situational awareness, you may be feeling surprised, angry, frustrated, or scared. I know. I felt all of those emotions and more.

Unconsciously Incompetent

I was surprised there were so many ways in which my situational awareness could be impacted and, consequently, my decision making could be flawed. I was angry that no one had taught me this stuff. I was surprised I'd not had anyone hurt as a result of what I didn't know. I served as a commander for 22 years and during that time commanded thousands of emergency incidents. The more I learned about situational awareness, the more I realized how many times, and in how many ways, my situational awareness had been impacted in the past. I was, no pun intended, clueless. I didn't know what I didn't know. After 20 years of commanding thousands of incidents, I found myself in the uncomfortable position of realizing I was unconsciously incompetent—I was unaware of all that I did not know.

This chapter summarizes some of the most important lessons I learned during my research, supported by numerous neuroscience and cognitive psychology studies and a review of hundreds of near-miss and casualty reports from a variety of professions where high-consequence decisions are made under stress. The lessons reflect the gift of knowledge shared by expert-level fireground commanders who agreed to allow me to study their situational awareness and their decision-making processes.

A number of these lessons come from research outside of emergency services. This is due, in part, to the funding available for research in other domains. The military, medicine, nuclear power, and aviation industries invest millions of dollars in research for the purpose of improving decision making. These industries, not unlike the emergency services, stand to suffer significant losses from poor decisions attributable to flawed situational awareness.

There is, literally, a whole world of research and researchers out there of which most first responders are unaware. In fact, when I stumbled into this other world I recall how surprised and fascinated I was that so much research had been done and I was completely unaware it. The lessons from research outside of public safety are vitally important to improving our understanding of high-risk, high-consequence situational awareness.

I truly feel it's my calling to share the lessons I have learned with first responders, and I feel blessed to have been given the opportunity to do that in both this book and in live presentations delivered across North America, Europe, and Asia. The many lessons I gained have been shared with you throughout this book; however, a few are so important I feel they warrant their own chapter in this book. Here they are.

Lesson 1

Responders with poor situational awareness can still have a good outcome, if only by luck.

The reason this lesson is so critically important is because when people make good decisions (by luck) while operating with flawed situational awareness, they may never know their situational awareness was flawed in the first place. The lack of a poor outcome may give responders the confidence to think their situational awareness was strong and that led to the good decision when, in fact, they were only lucky.

When the outcome is good, it may give responders the false belief they had strong situational awareness throughout the entire incident. This deceptive outcome can give decision makers a false sense of confidence to continue to do things in the future the same way they have been done in the past, and responders may be completely blind to the fact that their situational awareness is flawed because there has not been a bad outcome.

Lesson 2

A decision made with good situational awareness can still have a bad outcome.

Not every bad outcome can be linked to a problem with situational awareness (though if you dig deep enough, there's often a connection). Because emergency service providers work in a highly unpredictable, unstable environment, unexpected things can happen, and that can lead to bad outcomes.

One example that comes to mind is an incident I was told about where a building was booby trapped and set on fire by an activist group who was trying to disrupt genetic animal research. The firefighters entered the burning building having no idea that it had previously been broken into and the traps that lay in wait for them. The interior crew opened a door, which resulted in the purposeful overturning of a container of gasoline intended by the activists to ignite and burn the firefighters. Neither the firefighters nor their commanding officers had any idea that such a trap was set with purposeful intent of harming responders. Fortunately, the trap didn't worked as planned. As the firefighters opened the door, the gasoline spilled onto the floor, but for whatever reason, it did not ignite. The firefighters found themselves slipping and sliding on the product on the floor and withdrew from the building. Upon their exit, personnel nearby who were not wearing breathing apparatus immediately noted the strong odor of gasoline that had saturated the crew's turnout gear. The crews and commanders were completely unaware of the trap. Had the gasoline ignited, the lives of the responders would have been imperiled.

The potential for the poor outcome at this incident could not be attributed to flawed situational awareness or poor decision making (figure 18–1). There were no previously identifiable clues or cues that the building was a strong target for terroristic activity. Additionally, there were no signs the building had sustained forced entry.

Fig. 18–1. A commander can do all the right things to maintain strong situational awareness (e.g., maintain a big-picture view, stay out of the action, and use a command aide) and still have a bad outcome. Not *all* bad outcomes are a result of flaws in situational awareness.

Lesson 3

Maintaining situational awareness requires a physical, mental, and emotional commitment to paying attention.

In the context of this lesson, paying attention means capturing clues and cues (Level 1 situational awareness), processing those clues and cues into meaning in the current environment (Level 2 situational awareness), and making predictions of future events through mental modeling (Level 3 situational awareness). It means being consciously aware that you are taking all the necessary steps to develop and maintain your situational awareness. In neuroscience parlance, the term is *meta-awareness*. Are you paying attention to the big picture? Are you seeing the most important things? Do you understand what they mean? Are you able to, within your mind, look into the future and see what the incident should look like at some point later in time (for example, 5 or 10 minutes out)?

Because there are so many other things that can cause distractions and interruptions and draw your attention away from the important task of maintaining situational awareness, it's absolutely critical to conduct self check-ups. Otherwise, your situational awareness can slip away so slowly and subtly that you won't even realize you're losing it (figure 18–2). Then you'll have something catastrophic happen and wonder how you didn't see it coming.

Fig. 18–2. Commanding an emergency in a dynamically changing environment requires a physical, mental, and emotional commitment to paying attention.

Lesson 4

What you should pay attention to is not always intuitive or obvious.

This is where your primal instincts can be a hindrance instead of an asset. You may recall from chapter 11, where we discussed attention-management barriers, that under stress your attention can be drawn toward things that are loud, bright, moving, and/or in close proximity to you. Adding to that, your attention may also become fixed on what you perceive is the most important clue or cue at the incident (e.g., flames or the shooter with a gun trained on a hostage). This narrowing of attention can impact your situational awareness, and you may not consciously know it's happening to you.

I have demonstrated this numerous times in exercises involving first responders. I put a photograph on a screen and tell the participants that something on the screen is going to change. I ask the participants to pay attention and when they see it, write down what changed. It's not uncommon at all for almost everyone to miss the change because it happens in a subtle way (figure 18–3).

After two or three times showing the picture with only a handful picking up on the change, I tell the audience that I will reveal the change by pointing at it with the cursor. Then I take my mouse and I move it up and down quickly, all the while pointing to what is changing. With each movement of the mouse, the tip of the cursor touches what is changing on the screen. When the demonstration is completed, I ask if everyone saw what changed, and most giggle at the seemingly rhetorical question.

I'm sure many are thinking, "Who could *not* see it change? You were pointing your cursor right at it!" Then, to the amazement of the audience, I ask for the participants to be honest with themselves and with the others in the room and raise their hands if they did not see what changed. In many cases one-third to one-half of the hands go up. Not only did these individuals miss the change the first three times I showed it, they missed it after I told them I was going to point my cursor at the change to *ensure* they would see it. Yet many do not see it. Why?

Fig. 18–3. What a commander should pay attention to at an emergency scene is not always intuitive or obvious. This requires a commander to be alert for subtle clues and cues that help build strong situational awareness.

The answer lies in their primal instincts. Remember your cave dwelling ancestors I talked about in chapter 3? Their primal instincts taught them that things moving toward them can be dangerous. Moving things can potentially kill you and eat you. So you'd better pay attention to the moving things. And so we do. We pay attention to moving things, like cursors. You might think it ridiculous to make this link; there's nothing whatsoever dangerous about a moving cursor on a computer screen. It's true, there isn't, so long as you are thinking rationally. Unfortunately, as you may also recall from chapter 3, the stressed brain does not think rationally. It thinks intuitively, and the way it does that, in part, is by tapping into intuition.

Contained in your intuitive programming is genetically encoded instinct that tells your brain moving things are dangerous and, therefore, should be watched. Oh my! You're more vulnerable than perhaps you ever thought possible. At emergency scenes there tend to be lots of moving things. Some of those moving things can be dangerous, but many of them are not. For example, a fellow responder coming up to talk to you is not a threat; however, your brain doesn't know that. In primal parlance, that responder might as well be a saber-toothed tiger coming in for the kill. You're instinct is to pay attention to this person even at the detriment of other very important things happening around you.

Loud noises, bright flashing lights, and fast moving things at emergency scenes are naturally distracting—natural because your primal instincts draw your attention to them.

Here's an example of how your situational awareness many be innocently stripped away from you. You're operating at an emergency scene, and one of your subordinates approaches you to have a discussion. The closer the physical proximity of the subordinate to you, the more you train your attention on that person. Remember, this is a primal reaction. Even though you know the person well, your instincts put you on guard because he or she might be approaching you with the intent of killing you and eating you. I know, it doesn't make sense. This person is a member of your department, but instinctively, under stress, your body is having reactions of which you are no longer in control.

When this happens, you may find your full attention being given to this person. So much so that if you had a radio blaring out messages from personnel operating in a high-hazard environment, you might cover up your portable radio microphone or turn the volume down so you can give your full attention to this threat standing next to you. I've done this myself, not realizing the potential consequences. Not only do I miss radio communications from my personnel who are in the highest-risk environment, but I also miss critical clues and cues that are essential to maintaining my situational awareness.

The potential for missing critical clues and cues results from my attention being given to someone who has walked up to me at the emergency scene. Wow! Think about how often this might happen at your emergency scenes.

Lesson 5

Responders rarely realize they're losing their situational awareness until it's too late.

Wouldn't it be nice to have some mechanism in your brain that would trigger an alarm to warn you that your situational awareness is eroding? Although some of the participants in my research reported they experienced an "*aha!*" moment—that sudden realization they were missing really important things and dialed back in—many have not.

> Some are fortunate to have had that epiphany where they thought to themselves, "I'm losing it here. I need to back up. I need to get control of this. I need to figure out where everybody's at and what they're doing."

Realizing the loss of situational awareness is not uncommon, but you may not always know it. Though I can't substantiate it with science, it's plausible to surmise that for some the intuitive *gut feeling* becomes so overwhelming that it serves as their alarm system.

I recently interviewed a commander with more than 40 years of experience, and he told about a tragic incident where two firefighters died while operating inside the building fire he was commanding. He said at one point during the fire a fellow command officer approached him and asked him how he thought things were going. He reported that his reply was short and to the point. "Not well. Something's wrong. I can't put my finger on it. It's not any one thing that I can tell you I see that's telling me, but something's wrong." Unfortunately, his intuition was sounding the alarm, but he didn't pay attention to it because he wasn't able to rationally justify why he was feeling like he was. He was reverting to rational judgment. He was probably thinking to himself, "If something truly is wrong, then I should be able to see it and describe it to my associate." Because he couldn't, he dismissed his intuition as happenstance, and it resulted in a tragic outcome.

For many of the participants in my research—and I think for many who have been in charge during incidents with tragic outcomes—their situational awareness had faded away and they were oblivious to the fact. They weren't capturing the most relevant information, they didn't comprehend the meaning of most important clues, and they weren't predicting future events through modeling. Thus, by definition, their situational awareness had eroded, and they never got the wake-up call until the catastrophic and significant near-miss event occurred.

Lesson 6

It's critically important to be able to form mental models of both the past and the future.

There are many, many clues and cues exhibited prior to your arrival on the scene of an emergency. If you're fortunate, some of these clues and cues are relayed by civilians through your dispatcher.

However, it's important to keep in mind that the same stress that impacts your situational awareness is also impacting civilian observers. Arguably, the civilian may have greater stress impacting their situational awareness than a responder, because civilians don't have the direct, first-hand experiences with emergency situations. It's much easier for civilians to misunderstand what they are seeing or for them to see and report something that appears to be very important yet, in reality, is not.

Equally, it can be extremely difficult for dispatchers to extract accurate and reliable information from civilian callers (figure 18–4). Having worked in a dispatcher center, I have experienced the frustration of having multiple callers reporting the same incident in multiple ways and trying to make sense out of what is really happening. Think about how many times you've responded to an incident that was not as bad as originally reported. Thus it may be difficult to rely on civilian information to help form your situational awareness prior to arriving at an incident.

Fig. 18–4. It can be extremely challenging for dispatchers to extract accurate and timely information from civilian callers who are upset.

Unfortunately, you don't have much choice. Your brain is going to process the information it receives through dispatch, and it's going to form mental models about what it perceives is happening. This isn't necessarily a bad thing. In fact, it's vital that you form some semblance in your mind of what is happening prior to your arrival. Certainly something is happening, or there's no emergency. Unless you are present at the inception of an emergency, you're coming into the incident well after its starting point. And rarely is an emergency incident static; rather, the incident conditions are dynamic and rapidly changing.

Taking a rearview mirror view of the incident is critical to forming situational awareness. You must form a mental model of what happened before you arrive. You need to determine how this situation got to where it is at the moment of your arrival.

Using a structure fire as an example, the building and/or its contents have burned for a while prior to your arrival. Perhaps it started as a very small fire and grew into the conditions you now see. Forming a mental model of what happened before your arrival enables you to draw on your tacit knowledge and form some expectation of how the incident is progressing.

If you arrive and see a fire that hasn't extended beyond the room of origin, this tells you something about how aggressively the fire is moving and how fire and smoke conditions are expanding. Similarly, if you arrive and see fire that has extended beyond the room of origin, this may tell you a very different story about the aggressiveness of the progression and the growth of fire and smoke conditions. These clues and cues give you an indication of the *speed* of the incident.

The speed at which the incident is moving is critically important information because you need to form a mental model based on the application of your tacit knowledge to the events that occurred prior to your arrival (figure 18–5). Then you have to make some decisions about what actions you're going to take with the resources (personnel, equipment, apparatus, water, and so on) you have at your immediate disposal. At this point, the critical question you must answer is: Can my resources overwhelm the incident, slow down, and eventually stop the progression of the undesirable events? If you are confident in the affirmative, you'll likely engage the enemy offensively. If you are not so confident your resources can outpace the progression of the incident, defense may be your best course of action while you wait for the arrival of additional resources (which may or may not arrive in time to allow you to overwhelm the incident).

If you deploy when you are under-resourced, you're taking grave chances with the lives of the responders in your charge. As I discuss throughout this book, there are numerous explanations for why a decision might be made to engage while under-resourced—ego, overconfidence, peer pressure, underestimating the speed of the incident, and overestimating the abilities of your personnel, just to name a few.

Fig. 18–5. Every incident has a speed. And, like a race car, it is impossible to tell its speed simply by looking at a snapshot. While an initial incident size-up may be a snapshot, the commander must also watch the incident move over a period of time to fully understand how fast conditions are changing.

You next need to form a mental model for predicting future events. This entails asking yourself three essential questions:

1. Where is this incident headed?
2. How long is it going to take to get there?
3. Can the resources I have at this moment change the course of the incident in a favorable way without jeopardizing any lives?

This is your look into the future—forming your future mental model of the outcome of the incident. As previously noted, you are vulnerable to biases and human factors that can skew your confidence levels. Those most at risk are the ones who believe every incident can be managed with aggressive tactics. This simply is not true.

The best example may be a military scenario. If every war was won with aggressive tactics, the side whose soldiers were most aggressive would always win and would do so most expeditiously. But we know this does not happen. Why? For one thing, the enemy's military may be bigger, stronger, and better provisioned than the opponent's. To rush into battle under such conditions would likely mean certain death and would do little to accomplish the objective.

The same can be said when responders arrive at an emergency and the enemy is bigger, stronger, and better provisioned than the arriving resources. Simply stated, a fire producing British Thermal Units (BTUs) at a rate greater than the water you apply can absorb is going to win every time. Additionally, when a building's structural components start to give way from the effects of heat, gravity is going to win every time—and the building is going to collapse.

The mental model of future events entails your ability to look at the clues and cues and *realistically* predict what the incident is going to look like at some point in the future (e.g., 5 or 10 minutes into the future). This prediction of the future includes the deployment of your resources by whatever tactical operation you decide. You are predicting whether the actions of your personnel are going to have a profound impact on slowing and stopping the progression of the incident.

In your mental model, you should envision what a successful outcome would look like. This prediction of the future becomes the expectation against which you compare your progress. As I write this I am reminded of a leadership principle taught by Stephen Covey in his best seller *Seven Habits of Highly Effective People*. That principle is: Begin with the end in mind. Applied to emergency responses, begin with your expectation of what the successful incident outcome is going to look like. Mentally model out the progression of the event to the successful outcome. Then—and this is critical—compare the actual progression of events to your mental model. If things are going well, they should match up. If things are not going well, they won't align, and this should set off major alarm bells in your head.

This alarm may come as explicit (conscious) knowledge that things aren't going well, or they may come more subtly as a gut feeling—your intuition (subconscious knowledge) alerting you that what is happening doesn't match your tacit knowledge (the mental library of your training and experiences). As noted previously, don't dismiss your gut feelings;

they are telling you something. If your intuition is telling you things are not going according to plan, look at the clues and cues and compare them with your expectations. Chances are very good that you're going to see they do not align.

As you arrive at the scene, not only should you use the clues and cues to predict the future, but you should also use the clues and cues to predict the past. How did this situation become what you see now? How did it deteriorate? Chances are good the incident hasn't got better with age. Instead, as the seconds and minutes passed, the situation was in a steady state of change. That change usually comes in the form of degrading conditions and elevated risk for the responders. How long did it take for the incident to degrade to its current state? How fast is the incident degrading at the moment you arrive? Can the resources you have at the current moment in time overwhelm (control or suppress) the hazard? It's critically important to understand the speed at which the incident is moving because you have to decide if your resources can go faster. If not, you're going to get overrun. You will almost always arrive at an incident initially understaffed. Underestimating the speed of the incident while overestimating the abilities of your resources can be a recipe for a catastrophe. Simply ask yourself: Can I accomplish the goals that I set with the resources that I have?

Five Sure Ways to Get Into Trouble

Because I have intensely studied near-miss reports, line-of-duty death investigations, and case studies, I have noticed there are some common themes. In this section I talk about five of those themes. There are certainly more than five, but these are the ones that jumped out at me from the reports. I call these *red flags*.

Red Flag #1

Failing to process the meaning of critical clues and cues.

If you're like me, when you read a case study of a near miss or a line-of-duty death, or watch a video where an incident is degrading, you see the critical clues and cues jumping off the page at you. They're all over the place. It's hard to imagine how the people operating at the incident did not, or could not, see them. If you've read the entire book up to this point you now understand there are many, many ways for responders to not see what's happening right in front of them. They're not missing the important clues and cues on purpose but they are, nonetheless, missing them.

Each type of emergency has its own list of five to seven most important critical clues. For a structure fire, one of the most important clues is the smoke. This wasn't taught to me when I was a new firefighter. We were taught to look for the fire; the smoke was just something that got in the way of what we were supposed to be looking for. It was treated like an annoyance. If not for all that black smoke, we'd be able to see the fire and *then* we'd know what was going on and what we need to do.

Perhaps back in my early days in the fire service we could get away with that. Houses were built differently, and the contents in those houses were different. Things were built sturdier. There were far less hydrocarbon-based construction materials and furnishings being used.

That's all changed now. The smoke that is generated from decomposing hydrocarbon-based products is fuel—as explosive and dangerous as gasoline. The color, volume, velocity, and density of smoke are critical clues. In many cases it's the smoke that indicates what's burning, where it's burning, how long it's been burning, how deep into the building it's burning, and whether the building is under pressure from the buildup of gases. To fail to comprehend the meaning of smoke is to miss critical clues and cues about what is happening.

Borrowing from the medical field, smoke might be one of the *vital signs* you want to be sure to capture. It may helpful to think of the building as a living being, and the smoke contains the vital signs—pulse, respirations, and blood pressure. The smoke can pulse, breathe, and create pressure. I'm going to stop there with the analogy because I don't want to infer that the clues and cues of smoke are the same as human vital signs. Just know that smoke is, indeed, a *vital sign*. If you miss the vital signs, you're far more likely to *misdiagnose* the root of the problem.

Red Flag #2

Underestimating the speed of the incident.

This is can be an easy trap to fall into. Recall from chapter 3 the recognition-primed decision-making process. The first step is to conduct a rapid *snapshot* size-up, not because the building will burn down if it's not done quickly, but because the clues and cues are changing so fast that if more time is taken to capture information, the incident can change to a completely different situation. So, for that reason alone, it must be a snapshot in time.

The downside of this is if you're only looking at a snapshot, it's difficult to comprehend changes in the clues and cues. Specifically, it can be challenging to determine the speed of the incident.

Think about looking at a snapshot of a car traveling down a highway. It would be very difficult, looking only at the photograph, to determine the speed of the car. However, if you were allowed to look at a movie clip of the car traveling down the highway, it would be much easier to estimate its speed in the context of how fast the car is passing by other objects.

Whereas the initial size-up may need to be quick, it's vitally important to watch the incident for a period of time to get a sense of how fast it's moving. It's the speed of the incident that helps determine if the resources available are sufficient to overwhelm and stop the progression. Tied closely to this is the next red flag—overestimating the abilities of personnel.

Red Flag #3

Overestimating the abilities of the personnel.

As unsettling as it may be to admit, every responder is *not* created equal. Differences in training, experience, age, fitness, and stamina are all factors that influence crew abilities. It's the same abilities that can weaken a strong crew or strengthen a weak crew—and every crew is, collectively, unique.

Making an assessment of crew ability is important to determining what they can accomplish in an established (and often finite) period of time. This is the very essence of Level 3 situational awareness—possessing the ability to predict future events through mental modeling. Recall from our discussion in chapter 13 that mental models are analogous to little movies that you play forward in your head to help predict the future of the incident. The *actors* in your movie are the personnel who work for you, with you, and around you. To project a crew's abilities to perform certain tasks in predictable amounts of time, it's essential to understand the characteristics that influence the amount of time it will take them to accomplish the task. Those characteristics are training, experience, age, fitness, and stamina coupled with crew size.

If there is an expectation that a crew can take a certain amount of time to complete a task, but that amount of time passes and the task is not complete, stress will likely be created. The person who made the prediction was expecting something to be done and that *something* factors into his or her Level 3 situational awareness.

This heightens the decision maker's stress if it causes worry about what may be happening to the crew that's keeping them from accomplishing the task in the estimated allotted time. Equally as stressful, it may cause the decision maker to become frustrated and angry, focusing attention on where the crew is and what they're doing—not out of concern for their well-being, but rather concern for how the crew is messing up the decision maker's strategy by not getting done what needed to be done in the allotted time. Unless you know your personnel very well, it can be challenging to make accurate estimates of their abilities.

I was told of an impromptu study where commanders were brought to a training facility and told they were going to have a crew of firefighters perform a specific set of tasks. The commanders were then asked to estimate how long it would take the tasks to be accomplished. Consistently, the commanders' estimates were only one-half to one-third the time it actually took the crews to accomplish the tasks. Some commanders were astonished at their overestimation of their crew's ability to perform the tasks. Other commanders blamed the crews and offered excuses for missing the estimation—completely missing the intended lesson.

When stress levels go up and personnel are operating on the scene of a high-consequence incident scene, there is a propensity to overestimate the abilities of crews because there is an inherent interest in seeing an expedient resolution to the situation. An inaccurate estimation of crew ability can flaw the entire mental model, and situational awareness will deteriorate.

Another unfortunate consequence that can occur when estimating the abilities of crews is to make a reasonably accurate projection of the crew's progress and find that, because of the conditions of the incident, the crew does not accomplish the task in the allotted time. Again, this isn't because the crew's ability was overestimated. It's because the incident conditions prohibited the crew from completing the task.

Let's say, for example, a decision maker sends a crew to perform a high-risk task in a hazardous environment and then makes a mental note about how much time it should take for that task to be completed. When that allotted time comes, the decision maker has a choice to make. One choice is to continue on the chosen course of action for the crew, allocating the crew some additional time to complete the task. Another choice is to commit additional resources to the task under the belief the resources initially committed to the task were inadequate. Alternatively, the decision maker may tell the crew to retreat from the hazard zone to a position of safety. Each of these choices has inherent risks.

Give the crew more time

Giving a crew more time in the environment may be based on the assumption the crew is making progress, but they just haven't reached the point yet where they have overwhelmed the enemy. If true, pulling the crew out prematurely may cause a greater loss than necessary. However, it's possible that the commander may, for a variety of reasons, fail to see the crew as being in a losing situation. These reasons can include ego, overconfidence, aggressive culture, Type-A competitive personality, and so on. The decision is made to give the crew additional time because the commander doesn't want to admit defeat against the enemy. To admit defeat takes a great deal of self-esteem, fortitude, and confidence. In fact, if a commander tells the crew to withdraw from the high-risk environment, the commander may to receive some contrary feedback (e.g., "Give us just a few more minutes, we're making progress!"). On numerous occasions I've been the recipient of this kind of feedback; I've also been the provider of this feedback.

Why would a crew raise such objections? A couple of explanations are plausible. The crew can be unaware that the conditions have changed and they're at-risk. The commander, operating outside the hazard zone, can see the big picture and can see clues and cues that are different from those operating in the thick of the action. Also, the same traits that can cause a commander to resist defeat make it difficult for the front-line personnel to admit defeat. The act of retreat is the admission of defeat, and that can hurt self-esteem.

Allocate more resources

Allocating more resources to the problem could be a viable tactical decision as long as the backup resources are adequate enough to overcome the situation. Sadly, sometimes this is not the case. In fact, the review of some casualty reports reveals no amount of resources available to the organization would have been able to overwhelm the situation. Dedicating additional resources to the problem, perhaps as a demonstration of defiance or refusal to admit defeat, only needlessly jeopardized the lives of the additional responders.

Withdraw

Ordering withdrawal of personnel from the high-risk environment may result in potentially undesirable, albeit non–life threatening consequences. In addition to personal self-esteem wounds that admission of defeat can cause, responders may turn against the commander and harshly criticize the decision maker as having no backbone or lacking in aggressiveness. If the withdrawal results in additional loss of property (as it often does), there is also a risk of additional criticism from those directly impacted by the loss—the property owner, community members, elected officials, media, and in some cases even members of other public safety agencies.

I recall an incident where a fire department responded to a well-involved house fire. Because of the extent of the fire and the lack of experience among the responders, the commander chose to be defensive. I wasn't at the scene of the fire, so I can't opine on the appropriateness of that strategic decision; however, when the neighboring mutual aid department arrived, their officers felt the fire could (and should) be fought using aggressive interior tactics. They were denied entry by the incident commander, and they raised objections. The commander stood his ground, and the building ended up being a total loss. The members of the mutual aid department were subsequently overheard calling the host agency "an outstanding fire department" but not as a term of endearment to their tactics. Their backhanded compliment really meant they were calling the department "an out-standing fire department," meaning the department members stand outside to fight their fires. That criticism can hurt.

Estimate crew abilities

If a commander or company officer overestimates the abilities of his or her personnel, it can also cause issues with situational awareness. Recall awareness is vulnerable to the effects of stress. If a crew is given an assignment and they are taking too long to get the job done, the commander or company officer may become stressed. That stress leads to confusion, frustration, and anger.

The commander may not understand why a crew is taking so much time to perform a task that should already be completed. This heightened concern for crew performance can draw the commander or company officer's attention to where the crew is located and what the crew is doing.

> While conducting my research, I had a commander admit: "I can get so frustrated I just want to give command to somebody else and go in and do the job myself, because then I know it'll get done—done right and done quick."

From reading casualty reports, I'm confident that is exactly what some commanders do. This is problematic. As discussed in chapter 9, when commanders go hands-on, they are no longer in a position to comprehend the big picture clues and cues and, by definition, are no longer commanding the incident.

> **Red Flag #4**
>
> **Feeling pressure to take heroic action without considering the risk and benefit.**

There has been considerable debate around the issue of conducting risk assessments and victim survivability profiles. My view on this is it relates to an unwillingness to graciously accept defeat against an enemy that's simply larger, stronger, and better equipped. First responders simply don't like to lose. I understand that. I don't like to lose, either. In fact, it's difficult to even write about losing because it gives me a sense of loss and makes me feel like a failure.

The medical profession has made better progress in this area than responders have. Doctors are trained right from the beginning to value human life and to do everything within their knowledge, power, and abilities to preserve it, protect it, and save it. Yet, the best doctors know when they have been defeated and talk to patients and/or their families about signing a Do Not Resuscitate (DNR) declaration. This is the admission of defeat to the enemy—in this case, a disease. The doctor knows there is nothing more that can be done, and the patient is allowed to expire with the administration of medication only to ensure his or her comfort. Life and death medical decisions are made all the time.

When I was new in the EMS field, DNR orders were not as common as they are today. Thus, I was forced (and *forced* is the right word) to resuscitate elderly persons who were simply at the end of their life cycles as aggressively as I would have resuscitated a 40-year-old cardiac-arrest patient.

In my early years in EMS, we also performed cardiopulmonary resuscitation on major trauma patients. Now many protocols prohibit it because the survivability of a cardiac-arrest patient, secondary to multisystem trauma, is so low.

Some emergency responders haven't gained a comfort level with conducting risk assessments and performing victim survivability profiles.

Can we ever be 100 percent certain that a victim is not somehow, miraculously, still alive in an environment that is obviously not survivable, even for fully protected firefighters? The obvious answer is no; we can never be that certain. In fact, such certainty rarely exists anywhere in nature.

All we can do is give thoughtful consideration to the probabilities, just like other professionals who deal with life-and-death situations. I acknowledge public safety providers are passionate care providers for the citizens we serve. That fact is not lost in my discussion. However, we must be equally passionate about ensuring we don't take undue risk when the probability of victim survivability is very low and/or the probability of a responder casualty is very high.

I've read too many line-of-duty death reports where firefighters died while conducting interior operations at structure fires when there were no victims inside. Or the valiant efforts of firefighters to save a confirmed victim resulted in the deaths of both firefighter(s) and the victim, only to have the autopsy reveal the victim had expired from carbon monoxide exposure prior to the fire ever reaching the victim.

This may be one of the most controversial topics I address in this book. The *go* or *no-go* decision is one of the hardest a responder ever makes, and one whose consequences the responder may have to live with for a very long time.

Until responders learn to make decisions based on the scientific evidence before them (like doctors do with DNR decisions), we are destined to rush (some might say foolishly) into situations with reckless abandon, and the outcomes continue to reflect a disdain for the laws of physics that should be both understood and applied to the decision-making process.

I made an argument in chapter 3 where I discussed the role of emotions in decision making. You may recall I noted emotions are essential in the decision-making process. Thus, to avoid the appearance of sounding contradictory, there's a place for both emotion and science in the *go* or *no-go* decision-making process.

Decision making becomes safer when the emotional component of a high-risk, high-consequence decision is balanced against the impact of losing a civilian or the consequence of losing a responder. Responders are under a lot of pressure to perform at an emergency scene, and some of that pressure may come from the department's own members manifested through the organizational culture.

Peer Pressure and Public Expectation

If the department's culture is one of *always* using aggressive, interior, all-out-all-the-time tactics, regardless of the conditions, the organization may have a higher propensity for a casualty. How can I make such a prediction? Because not every incident warrants such tactics. However, if members have only one tactical play seared into their subconscious minds from repetitive performance of aggressive tactics, they have formed . . . a habit.

Repetition in learning builds both cognitive memory ("What should I do?") and muscle memory ("How should I do it?"). Repetitive aggressive performance at emergency scenes and in training scenarios causes an aggressive script to be stored in the subconscious brain.

> When aggressive tactics are not accompanied by negative consequences (i.e., no one gets hurt or killed), it may lead to a false sense of confidence that such behavior is acceptable.

So it becomes the norm—the standard. This phenomenon is also known as *the standardization of deviance, drifting toward failure,* and *error creep* (see chapter 13, "Mission and Goals Barriers"). Regardless of the name, it is a real and present danger to first responders (figure 18–6).

Fig. 18–6. Peer pressure can have a strong influence on firefighter behavior. This is especially true of newer responders who want to "fit in" and be respected by their more senior counterparts. This may cause them to take risks beyond their capabilities.

In addition to the peer pressure that can come from the organizational culture, responders can also feel pressured to take heroic actions based on what they perceive to be the public's expectations of them. If responders get caught up in the notion of being heroes or saviors, they may find themselves feeling compelled to act in ways that affirms the public's perception as they believe it to be.

A responder who takes a defensive position in a high-risk scenario doesn't look nearly as heroic as the one who rushes in to save property. The responder who takes a defensive position isn't going to be the recipient of public accolades. How often do you hear of a public safety agency giving recognition to a provider who made the decision not to engage in a high-risk event with a high probability of a poor outcome? It would be rare—very rare. In fact, there are raging debates about whether responders who take undue risks should be rewarded for their actions, even if the outcome is good (which is often based on a combination of luck and skill).

Some argue it takes great courage to go against the odds and do heroic acts without regard to one's own safety. Others argue that to take undue risk is irresponsible, and to recognize and reward such acts is even more irresponsible. The argument is rewarding unwarranted risk provides positive reinforcement for undesirable behavior. There's a risk that someone else seeking the same accolades might take equal (or greater) risks in an effort to earn the admiration of peers, superiors, or the community (figure 18–7).

I'm not here to judge first responders. Rather, I'm simply sharing observations of some very basic traits of human behavior. Among the most notable, people *will* do things to get attention—good or bad; it's all about the attention. Add the macho persona of some responders who puff out their chests with aggression at the mere suggestion that tactics *not* be offensive, and you can quickly see ego, the need for attention, and the need to feel appreciated runs deep in the human psyche.

Some might argue that such thoughts are the farthest from their minds when they're called into action and lives are at stake. I agree, in part. It might be the farthest thought from your *conscious* mind; however, you're not in control of your *subconscious* mind, and such basic human needs are always going to factor into the subconscious influence on your behavior.

Fig. 18–7. Sometimes the presence of bystanders or distressed property owners making demands on the fire department for action can cause responders to take excessive risks.

> This is a risky profession, and I'm not advocating for creating work rules that eliminate the risk of your job. That's impossible. However, I do think there is a distinct difference between assuming the risk of an incident and creating the risk through reckless behavior.

Red Flag #5

Focusing on the wrong things or trying to process too much information.

You may recall from our discussion in chapter 4 that the number of unrelated pieces of information a person can recall is seven, give or take two. The heart of this issue is first responders are bombarded with information. It's easy to get overwhelmed if you try to remember and make sense of everything that's happening. Your brain simply does not have the capacity to do that.

One of the sad realities of this shortcoming is what your brain decides to remember may not be the most important pieces of information needed to make a quality decision. I discuss this in chapter 13 and have demonstrated this often in small groups where I tell someone a fact-laden story and then ask that person to retell the story aloud to everyone. Participating in this exercise causes the working (short-term) memory to be vulnerable to missing details. And the *details* that are forgotten can be some of the most important pieces of information in the story. Some tend do pretty well at remembering the very beginning of the story and the very end of the story, but the details in the middle tend to blur.

Now, when I say *blur* I don't necessarily mean they are forgotten. Sometimes the facts are confabulated—changed in order to make more sense in the mind of the participant. Although somewhat humorous (and maybe even mildly embarrassing at times), I have had participants leave out important details and make up stuff that never even happened.

All of this behavior is symptomatic of stress and data overload, and it's a real threat to public safety providers. It also happens in other high-stress, high-consequence work environments (e.g., medicine and military). It can even happen with witnesses to significant crimes (like murder). Two eyewitnesses to the same event can tell two different stories based, in part, on what pieces of data their brains perceived to be the most important (and thus are remembered) and the brain's uncanny ability to change the facts to form a coherent story they want to believe is the truth (even if it isn't). Such human behavioral traits can drive police investigators and criminal prosecutors crazy.

As a first responder you can improve your chances of capturing the right information through practice. In advance of emergency incidents identify and rehearse the act of gathering the essential pieces of information needed to make good decision. Because the information varies based on the type of incident, so should the rehearsal be varied.

Medical personnel are arguably some of the best-trained responders for this skill based on the number of opportunities to practice. If a victim is suspected to be having a heart attack, there's short list of critical information the provider gathers in order to make an accurate diagnosis. Likewise, the short list of information gathered for a suspected stroke patient may be different. The same is true for assessing emergency scenes. There's a short list of diagnostic indicators that provide a high probability the responder can make an accurate assessment of what is happening and what needs to be done.

Responders who try to process too much information are at risk of overload and this may cause their brains to shed information. As previously noted, sometimes the information that is shed can be vital information. Likewise, if you focus on the wrong pieces of information, this may lead to an inaccurate assessment of the situation. You're especially at risk of this happening when those on the scene make a really big deal out of something that isn't a really big deal. But because they are excited or upset about *their* big deal, their emotions can trigger your emotions and you may lock on to their information as being important, even if it's not. Just understanding you are vulnerable may be sufficient enough for you to consciously weigh the importance of data. As I noted earlier in this chapter, medical personnel do this all the time. If a patient is displaying the symptoms of a heart attack, the fact that he or she had complications from a knee replacement surgery 10 years prior is not going to be viewed as an important piece of medical data at that moment in time, and the provider is likely to dismiss it—at least initially while all attention is being focused on capturing the critical clues and cues and performing immediate life-saving treatments.

CHAPTER 19

FIREGROUND COMMAND BEST PRACTICES

As much time as I have spent over the last 18 chapters discussing the issues with situational awareness and its role in making good and bad decisions, this may be the most important chapter in the book. Arguably I was tempted to call this the "How to Fix It All" chapter, but I resisted because I'm not convinced the ways to fix all of these issues can be boiled down to a list of 10 action items. Nor am I conceited enough to think, despite my many years of intense study of this subject, that I understand this topic in sufficient depth to proffer such a comprehensive list.

That said, I am convinced that focusing on the best practices identified in this chapter will have you well on your way to reducing casualty incidents as they relate to situational awareness issues. And, perhaps the best thing about this chapter is . . . most of the best practices I recommend to you cost little or nothing. So no one can hide behind the we-can't-afford-to-make-the-recommended-changes excuse. Sorry!

Best Practice #1

Prioritize your incoming information.

You remember how many unrelated pieces of information a person can remember in short-term memory, right? (Seven.) Good for you!

Now erase that from your memory. When researchers put subjects under stress, the number dropped to five. Stress impacts your ability to hold onto and understand information—especially complex information. For the sake of providing a tangible solution, I address what I feel are perhaps among the more important pieces of data needed to comprehend what's happening at a residential dwelling fire. Other information is just as important; I'm simply saying that you may struggle to capture, comprehend, and recall more than five pieces. To give you some margin of safety, I give you four.

Critical data for residential dwelling fire decision making:

1. Smoke and fire conditions.
2. The construction and decomposition of the structure.
3. The speed the incident is moving.
4. A realistic assessment of savable lives.

Let me spend a little time with each of these, sharing with you why I feel the items on this list are so important. Arguably these four pieces of information aren't the only data needed in the decision-making process at a residential structure fire. But, if you can keep this data on the forefront of your mind and decision-making process, you'll be well served.

Smoke and Fire Conditions

When I was going through my basic firefighter training we focused on fire, not smoke. In fact, smoke just seemed to be an inconvenient obstacle that slowed our access to the heart of the issue—the fire. I remember my drill instructors saying, "Don't ever spray your water on the smoke. You wait until you can see the fire and then open the nozzle." I didn't question their instructions. Water sprayed on the fire made sense to me. Water sprayed on the smoke didn't. I credit Dave Dodson and his "Art of Reading Smoke" class for changing my entire perspective on this. Although fire is important, smoke is even more important.

The fire tells you what's burning at a current moment in time. It doesn't take much situational awareness to look at flames blowing out a window to decide the room is on fire and it's damn hot in there. The smoke, on the other hand, tells you something completely different. As Dodson notes, the color, volume, velocity and density of the smoke can tell you what's burning, where it's burning, how deep into the building the fire is, how much pressure the smoke is putting on the building and, most importantly, where the fire is headed. There you have it! Smoke, not fire, is the predictor of the future. Those four attributes of smoke serve as the primer for Level 3 situational awareness. It's far beyond the scope and intentions of this book to teach you how to read smoke. For that, look to Dave Dodson's "Art of Reading Smoke" program. It's on my list of essential training programs for firefighters.

I have watched videos where firefighters had to bail out of windows because of flashover conditions. Commanders' interviews of these incidents often contain quotes such as: "The fire conditions changed very rapidly and unexpectedly." I give them credit for half the statement being true. The *fire* conditions did change very rapidly. But the *unexpectedly* part is suspect. If you know what to look for, you understand the building construction and the physics of smoke and fire, and your situational awareness is strong, the smoke gives you clues and cues that flashover conditions are developing.

The challenge is seeing it, knowing what it means, and then, most importantly, taking action to withdraw firefighters in time to avoid their exposure to the massive fireball that the flashover is going to produce. If you're a student of the casualty reports, you know as I do that this lesson has not been learned in many jurisdictions. Smoke and fire follow the laws of physics. They are not *suggestions* of physics; they are *laws*. And they're predictable laws. There are very few surprises in the application of physics in the real world, and the way smoke looks and behaves prior to flashover is often both observable and predictable—if you know what you're looking for and you can see it.

Construction and Decomposition of the Structure

To understand how a building behaves when it's on fire, you must first understand how the building is built. This requires knowledge of building construction.

> I credit the late Francis Brannigan for helping me understand the critical importance of building construction. I recall him saying, "Every building has a common enemy and that enemy is gravity."

Indeed, Brannigan was right. Like smoke and fire, building behavior also follows the laws of physics. It is critically important to understand how buildings are constructed and how they behave when the structural components are weakened by fire.

Gravity is trying to make every building fall down—even those not on fire. Eventually, left unchecked, a building on fire weakens and gravity wins its battle to bring it down.

This lesson is catastrophically important—today more than ever. Why? Lightweight construction has changed all that we know and have come to expect of the behavior of buildings on fire. Cheaper, thinner, lighter (glued-together) building materials have replaced more expensive, thicker, heavier, and durable materials. The outcome has arguably been great for the profitability of contractors but not so great for the safety of firefighters.

Look at the line-of-duty death reports and you see the trend. More and more firefighters are being killed in structural collapse incidents in homes with lightweight construction. Until we change our tactics (or homes are outfitted with residential sprinklers), you can expect this trend to continue.

The Speed the Incident Is Moving

Every incident has a speed. This speed is usually assessed in the terms of how fast conditions are degrading and how quickly the incident is slipping out of the control of the responders. Arguably, some incidents are so fast that responders can't get ahead of what's happening. A fast-moving wildland fire comes to mind for me as a great example. A team of firefighters is not going to stop a 30-foot-high wall of fire traveling up a canyon at six miles per hour. And every incident has a pace. For some hazardous materials calls, the pace may be very slow so long as the product is contained and not presenting an imminent threat to lives or the environment. The speed of an incident in which a deranged person barricades himself in his home (with no hostages) moves at a slower pace than a bank robbery/hostage situation where there is imminent danger to the lives of civilians.

Part of the responder's mental modeling process is to assess this speed of the incident and then determine whether responders can outmaneuver (i.e., be quicker than) the pace of the incident's progression. This requires someone—maybe more than one someone—to look at the incident over a continuum of time.

This advice runs counter to the rapid size-up that is step 1 in the recognition-primed (RPD) decision-making process. RPD requires the rapid assessment because of how fast conditions can change; the decision maker must lock on to some point of reference in order to make rapid decisions. However, it's also vital for the decision maker to look at the incident over a period of time (e.g., several minutes) to conclude how fast the situation is moving.

Think of it this way. The rapid RPD size-up is the equivalent of a snapshot in time. The longer assessment used to determine the speed of the incident is the equivalent of watching a movie. Looking at an incident over a period of time gives you the ability to make the same assessment and then allows you to gauge your potential success, based on the resources you have at your disposal at that moment in time, to overwhelm the incident.

Realistic Assessment of Savable Lives

The harsh reality is that not all victims are savable. Some become mortal casualties prior to the arrival of responders.

> In fast-paced environments where rapidly changing conditions impact victim survivability, there is often a narrow window of time in which a responder can actually influence the outcome in a positive way.

I refer to these continuums of time as *windows of opportunity*, and there are essentially three of them. The first is the window for *civilian* survivability, the second is the window for *responder* survivability, and the third is the window of *structural* survivability.

Victim survivability

Skin begins to melt at around 160 degrees. If you think about it, that's not very hot. The conditions inside a structure fire can easily and quickly exceed 160 degrees. Firefighters know that, in most cases, it's not the fire that kills civilians; it's the smoke and the heat contained within the smoke. Oftentimes by the time the fire reaches the victims, they have already perished from the inhalation of superheated gases or exposure to hot gases.

The window for victim survivability in the harsh conditions inside a structure fire is very narrow. Fortunately, most victims self-extricate from buildings on fire. They are alerted by fire alarms or the smell of smoke and exit the structure prior to first responders arriving. Those who were not alerted or who are unable to self-extricate (e.g., small children, elderly, and disabled) become the victims that firefighters are called into action to save.

These victims are savable only as long as the conditions inside the structure remain compatible with life. A bedroom full of thick, black, super-heated gasses is ripe for flashover and not an environment compatible with human life. Even if firefighters perform heroic deeds by entering the environment and extricating the unconscious, badly burned victim, the probability of victim survivability is nil. Simply talk to the medical personnel who work at a burn center for confirmation. Second- and third-degree burns over 70 to 80% of a person's body are not injuries compatible with life. If the victim is not dead on removal, it is likely the victim will expire within a short time frame. After the window of civilian victim survivability is closed, there's nothing firefighters can do to change the outcome. There's going to be a civilian fatality.

Firefighter survivability

The second window is that of firefighter survivability. There are certain fire conditions into which first responders can enter and survive with all of their protective gear on. When a fire is young and smoke and heat conditions are relatively mild, the window for firefighter survivability is wide open, and responders are readily able to enter the structure and perform rescue and fire attack operations. However, as conditions deteriorate and the smoke conditions develop to be hotter and thicker, the window of firefighter survivability begins to close. Protective gear allows the window for firefighter survivability to remain open longer than for civilian victims. But make no mistake about it, firefighter survivability in a structure fire also falls within a finite time frame. And when that window closes, little can be done to change the outcome of a firefighter casualty.

Structural survivability

The third, and final, window is that of structural survivability (figure 19–1). Similar to civilians and firefighters, the structure can only take so much before it dies (collapses). And like firefighters, the building *may* have protective equipment designed to help it survive the ravages of fire (e.g., fire resistive construction and sprinkler systems). The same way fire can overcome civilians and firefighters, it can also overwhelm systems designed to keep the building safe. When this happens, the building is going to lose, and it's going to fall down.

Fig. 19–1. It is a situational awareness best practice to consider the critical clues and cues (e.g., smoke and fire conditions, building decomposition, speed of the incident, and assessment of savable lives).

Extricating civilians and fellow firefighters from advanced, fast moving fires is considered a major challenge on the abilities of responders. Conducting the same types of rescues under the same advanced and fast moving fire conditions, with the additional complexity of structural collapse looming, is statistically and practically impossible in nearly all cases.

Best Practice #2

Set the strategy and tactics based on the quantity and quality of your resources.

Although this best practice seems like such a simple recommendation, putting it into action can prove challenging. Nearly every personnel-intensive incident is going to start from a position of being understaffed and under-resourced. Regardless of how many personnel you have on duty, or whether they come from home or from neighboring jurisdictions, operations almost always begin with the arrival of a single unit or company. In the case of police, that may be one person. For emergency medical services (EMS), it may mean a crew of two. For fire departments, the initial staffing varies widely but will likely range from two to five on the first arriving apparatus. In many cases, the emergency can be handled effectively with the initial response crew.

The problem starts when responders arrive, and the speed and complexity of the incident is beyond the capacity of the initial response resources.

Many first responders are, by their nature, problem solvers with Type-A, aggressive, competitive personalities. These are not bad qualities to have in first responders. In fact, arguably, they are the very qualities that allow first responders to be so effective.

However, combine a competitive, can-do, problem-solving disposition with incident conditions that are beyond the capability of the initial response crew, and you can anticipate the potential for problems. These problems are what I see when I evaluate casualty incidents—problems in the form of responders trying to defeat an enemy for which they are woefully under-prepared to defeat. The speed and complexity of the incident are no match for the resources present at the current moment in time.

Unfortunately, being outmatched is going undetected, under-detected, or ignored, and responders enter situations where the consequences, in hindsight, seem obvious and predictable. There's a lot to be said for rapid action as a factor for saving lives and reducing property loss. In fact, most first responder training and preparation is based on the rapid action model. I don't want the message to get misconstrued that I am opposed to the quick actions that often result in successful outcomes.

The problem occurs when responders repeatedly take excessive risks at emergency scenes. Over time, this high-risk, high-consequence behavior becomes routine and a scripted habit in the subconscious mind. For example, if your subconscious programming for the last 30 incidents has been rapid action that resulted in good outcomes, you become prone to rapid action as your default (scripted) response. This may occur at an incident where such action is against an enemy that far outmatches your capabilities.

You might speculate that the commanders who oversee these operations arrive on scene, see that their initial crew is overwhelmed, and order a quick retreat. Unfortunately, there are many cases where this doesn't happen, and the crews are left engaged in the fight against an enemy that's on its way to victory.

So why wouldn't the commander change the course of action?

Great question, and there are many plausible explanations including a loose application of Newton's first law of motion—an object in motion will stay in motion unless acted upon by an unbalanced force. As unfortunate as it is, when an action plan starts, oftentimes it's difficult to change. The commander may be blind to the unbalanced force (the fire).

Sometimes it's because the clues and cues are changing so slowly that the commander may not be able to see the changes. Sometimes it's because of overconfidence in the plan and the belief that just a few more minutes of effort will somehow reverse the deteriorating plan. Sometimes it's ego—simply not being able to admit defeat to the enemy and retreat to a defensive position.

All of these are formidable challenges, yet perhaps one of the most basic barriers to situational awareness has nothing to do with any of the aforementioned issues. Stated basically, when a crew is operating in an understaffed position, and a commander, safety officer, or operations officer is tasked with ensuring the safety of the crew operating in the high-risk, high-consequence, fast-paced environment, it creates stress. Based on everything you know as a public safety provider, I don't think you are be too surprised by that news. It's what happens when the stress level goes up—it causes situational awareness problems.

Stress, as noted throughout this book, causes the release of chemicals—neurotransmitters—that, in turn, change your behavior (based in part on how those chemicals trigger your genetically engineered, hereditary stress reaction). When this happens, people who are tasked with looking out for the safety of personnel operating in the hazard zone may find their attention narrowing down to the location and task being performed by the personnel most in danger. This is sometimes called *tunnel vision*, but in reality all the senses can become tunneled as stress increases. The senses tune into whatever is causing the greatest concern (stress). In this case, it's the location and activities of the responders who are in harm's way.

As the attention narrows, the ability to see and hear clues and cues in the periphery can diminish. This means commanders may not see and hear the very clues and cues that indicate the incident is degrading rapidly. If you have ever watched raw video of rapidly deteriorating emergency scene, you may have wondered how you can see and hear clues and cues to which crews operating at the scene seems to be blind and deaf.

When you couple the narrowing of attention with the impacts of hypervigilance, the commander may quite literally be deaf and blind to changing conditions. Recall that hypervigilance (again, triggered by chemical releases in the stress brain) puts the senses on alert for danger. That high level of alertness in a rapidly changing environment can spell disaster. Because the brain can take in, process, comprehend, and recall only five to seven piece of unrelated information, you can easily see how a stressed, hypervigilant brain can become overwhelmed. When overwhelmed with data, the brain can be slow to uptake and process new information. In fact, the brain can completely disregard new stimuli when it's reached its capacity for processing. Meaning, quite literally, stress can make you deaf and blind.

Commanders who miss seeing or hearing critical clues and cues may not be quite as incompetent as they are often labeled to be. This is not to say that commanders who miss critical clues and cues because they are not doing their job as a commander (e.g., they're hands-on engaged in operational activities instead of maintaining a big-picture view) are not incompetent. A good argument can be made for the incompetency of an individual who has the primary duty of ensuring the safety of personnel and fails to perform the essential task of maintaining a hands-off, big-picture view of the overall incident scene. Likewise, a strong defense can be made that a commander who is hands-off, watching the big picture, listening to critical radio traffic, and who still misses critical clues and cues is not incompetent—just human, with all the vulnerabilities that comes with the package.

If the commander arrives ahead of the operational forces, there's an additional risk I need to point out. It's possible that a commander will arrive, conduct a complete and thorough size-up of the situation, and develop a mental plan of action in advance the arrival of the tactical resources (see chapter 13's discussion on *mental models*). When this occurs, the commander may be at risk of developing and locking onto an action plan that, by the time resources arrive and get ready to deploy, is stale. The incident has progressed beyond the commander's mental plan. Yet, unless the commander starts the size-up process over again and develops a new mental model based on the status of the situation after the arrival of the tactical resources, there is a tremendous risk of putting a stale plan into action—a plan that won't work because the incident has progressed and the commander's original plan is no longer viable. Again, after a plan is formed and the decisions are made, it can be difficult to reconsider those plans. Not impossible; just difficult. The lesson here is to remember to conduct a fresh size-up and update the plan when resources arrive and it's determined that those resources are sufficient to overwhelm the enemy. If they're not, changing the plan prior to implementation is far easier.

When staffing levels (and the ancillary resources that support the staffing, such as water, pumping capacity, and so on) are not sufficient to overwhelm the enemy, the strategy needs to be defensive. I suspect there may be some readers who are turned off by that statement because of all the previously stated human qualities that make us good first responders (aggressive, Type-A, competitive, problem-solving, personalities).

The personnel operating in the hazard zone are most at risk and may only get one chance to communicate information if conditions change suddenly. The person in charge, or a designee, must ensure that operational radio traffic is monitored.

> Until responders are comfortable with the admission of defeat to enemies that are larger and faster, they are destined to continue on a path of undesirable, yet predictable, outcomes.

Best Practice #3

Never miss radio communications from your most at-risk personnel.

The act of listening requires a substantial amount of cognitive horsepower. We understand audible information by forming visual images in the brain. The audible cortex is connected to the visual cortex. Sparing you all the complexity, when someone shares audible information it's processed into visual images. It takes quite a bit of attentional energy to be an active listener. A commander who's trying to multitask by listening to the radio while watching the incident . . . while writing things down . . . while talking to the command aide or operations officer . . . while getting an update on the cell phone from dispatch . . . is going to suffer cognitive overload, and something is going to get missed (figure 19–2).

Fig. 19–2. It is a situational awareness best practice for an incident commander to have an aide to monitor radio traffic and take notes. This ensures that the commander does not miss critical clues and cues when having face-to-face discussions with members of the command team.

Unfortunately, the executive control center of the brain is going to make the decision of what gets missed. It won't be a conscious act on the part of the commander. Sometimes what gets missed is critical radio traffic, such as a Mayday that goes unheard. Sometimes it's critical changes in incident conditions that are communicated but not processed into meaning by the commander. For example, a crew inside a building radios, "We are on the second floor and conditions have changed and we have zero visibility." Setting aside the possibility the commander may be a novice (and that's not unrealistic in some organizations), a commander is not only susceptible to not hearing radio traffic under stress, but may also be unable to comprehend the meaning of radio traffic that is heard.

In addition to the impacts of sensory overload, the insidious affects of stress may also cause your brain to search for the simple solutions to problems (revolving mostly around the hereditary instinct of survival). In fact, under stress, you may find yourself not being able to make sense of complex information and situations. This may, in turn, cause you to become frustrated, confused, and maybe even disoriented.

Complex information and details are not a stressed brain's friend.

There are several viable solutions to these barriers rooted in communications. The first solution is to have someone dedicated to the act of listening to the radio. Remember all those military movies you've watched where someone's job is to carry and listen to the radio? There's wisdom in that strategy. The cognitive energy required to effectively listen and comprehend is tremendous, and does not lend itself well to a multitasked environment. The stressed brain becomes overloaded and then data gets shed. Unfortunately, it may be critical radio traffic that contained vital clues and cues to changing conditions.

Too, it's critically important for organizations to develop and use disciplined radio protocols so only the most important information is transmitted over the air. Too much radio traffic impacts everyone who hears it. Your brain can't *presort* which radio traffic is important and which is not. Thus, when traffic is transmitted, everyone who hears it has to engage in the uptake process of converting the audible information into visual images. If the airwaves are jammed with radio traffic, everyone who hears the traffic is vulnerable to overload and the shedding of data. The data shed may be different for each listener, which can compound the problem.

Radio procedures and disciplines ensure that only the most important information is shared. For example, radio traffic as innocent as "Tanker 9 is on the scene. Command, where would you like us to stage?" can create cognitive overload. In this particular situation, it's even more complex because the radio traffic is a question that requires a lot of cognitive horsepower. The listener has to process the question and then develop a visual image of the solution. Then the listener must form the thought of how to verbalize the response. And finally, the response must be formed into words and spoken over the radio.

> Load up a commander with just five to seven questions, and that may be enough to set him or her on the fast track to overload.

How do you avoid overload? One solution is to develop and use standard operating procedures (SOPs) along with radio discipline. One standard operating procedure might stipulate the commander establish the location of staging or water supply as part of the initial incident setup. Perhaps even better is an SOP that establishes guidelines for staging, water supply, and pre-incident plans that flag these locations so incoming units (presumably well versed on the SOP) can stage and establish water supplies in predictable ways based on established practices.

Another example of needless overload on a commander occurs when dispatch is contacted with a request for additional resources via a *single call* system. For example, the commander in charge of a fast-moving structural fire incident might need to summon multiple additional resources to the scene.

If the system set up to request additional resources is not part of a pre-established system, like the popular mutual aid box alarm system (MABAS) founded in Illinois, then the commander is forced to focus a significant amount of cognitive resources to think about which communities have the needed resources. The commander then communicates the requests to the dispatcher: "Dispatch from Main Street Command, please send me an engine from Johnsonville, a ladder from Smithfield, a tanker from Richland, and a tanker from Brownsville." But in a predetermined mutual aid system, the commander simply says, "Dispatch, give me a second alarm to Main Street and stage them at Fifth and Vine." This triggers the response of the same resources, but the heavy cognitive lifting that determined those resource needs was done in an office setting months or years previous to the actual fire event.

When it comes to radio communications it's important to be aware of your brain's limited cognitive capacity to uptake, process, and comprehend the information (with the duly-noted challenge that the more complex the information, the more challenging it can become). The more that can be done to prevent the commander from having to think about, prepare, and communicate complex information, the less vulnerable the commander is to overload (at least from communications issues).

At the heart of managing communication is the need to ensure the commander's ability to monitor the communications of the most at-risk personnel at the emergency scene. Think of the most at-risk as those operational personnel in the greatest danger and who, if in trouble, might get only one chance to transmit a distress signal to the commander.

> Missed radio traffic from at-risk personnel qualifies as a near-miss event, yet it is often overlooked or dismissed because there was no consequence when radio traffic is missed.

Here's one way to see if your organization is vulnerable to this situational awareness barrier. Following your next significant incident (i.e., one where personnel operated in a high-hazard environment), listen to an audio recording of the radio traffic. If you hear operational personnel calling command multiple times before receiving a response, you had a near-miss event.

For example, let's say following a residential dwelling fire you conduct a post-incident evaluation of the call. Listening the audio, you hear:

> T1: "Truck One to command."
>
> T1: "Truck One to command."
>
> T1: "Truck One to Command."
>
> C: "Command. Go ahead Truck One."
>
> T1: "We're on the second floor conducting overhaul. The crew is getting a little tired and we'd like to come out and go to rehab. Do you have any companies in staging that you can send in to relieve us?"
>
> C: "Affirmative, Truck One. I have several companies in staging. I'll send a crew in to relieve you, stand by."

This is an example of a near miss.

I use this example often when talking with first responders, and I often see the perplexed looks on the faces as I describe that radio traffic as a near-miss event. Some will even ask: "How could that be a near miss? No one got hurt. In fact, no one even came close to getting hurt."

And they're right. No one came close to getting hurt. This mind-set underlies one of the most disappointing misunderstandings about near-miss events: that someone had to get hurt or come very close to injury.

> Responders think that for an event to qualify as a near-miss, someone had to either get hurt or come so close to getting hurt that it's characterized as merely *luck* that saved them.

When you start to think of near-miss events on the basis of their untapped potential to result in catastrophe, you quickly realize it doesn't take looming consequence to qualify. All it takes is thinking through what happened while wearing a pessimistic lens. What if Truck One was in a tight situation and had only one opportunity to call command with a critical distress message. It's conceivable that command would not have heard that message either because their attention was drawn to something else or someone else. One of command's critical tasks (monitoring radio traffic of at-risk personnel) was interrupted by something.

If you are fortunate enough to see this event as a near miss—and many organizations are not—you would set out to understand what command was doing at the moment in time that Truck One called on the radio and command missed the traffic. Then set out to find a way to ensure it never happens again.

Either provide the proper support to the commander so that attention isn't drawn away to tasks of lesser significance or, if the commander was performing a critical command task, ensure there is someone else available to monitor the radio traffic.

When I talk to supervisors and commanders about this, they readily acknowledge this type of near-miss event happens in their departments all the time. At that point, I point out the need to take steps to ensure that this action is no longer condoned.

Because commanders are fallible human beings, their attention can be drawn away from monitoring radio traffic in innocent ways. For example, if a commander has a radio clipped on the turnout coat lapel and the microphone is mere inches from the commander's ear, it's nearly inconceivable that critical radio traffic would be missed, right? Well, not necessarily. Recall that hereditary instincts draw attention to occurrences that are loud, bright, moving, and within close proximity to the individual. Couple this with the propensity for tunneled senses in high-stress situations, and you have a combination that can set the commander up for failure.

Let's say the person is operating as a street-level commander and has a radio pinned to the lapel. The commander is listening to the radio traffic of at-risk personnel and is not missing any of the transmissions. Then, innocently, up walks the safety officer for a face-to-face discussion on the progress of the incident. At this moment, the commander can be faced with two barriers to situational awareness. The safety officer is moving toward the commander, which will, by instinct, attract the commander's attention because the brain doesn't know if this creature represents a lethal foe. I know it sounds odd for your brain to respond in this way to someone you know well, but it's part of your ingrained genetic stress reaction.

Also, because the safety officer's message is coming from a live person standing within feet of the commander, the message of the safety officer may take attentional precedence over the messages over the radio. At this point, the commander may, completely inadvertently, tune out the radio traffic because the voice over the radio doesn't present the same threat to the commander as the live person does standing within striking distance.

Equally consequential, a commander may put his or her hand over the radio speaker to muffle the sound while having a discussion with someone. Or, perhaps worst of all, the commander may turn down the radio volume. Why is this one worst of all? Because in a high-risk, high-consequence, fast-paced environment, the commander may simply forget the radio volume was turned down and subsequently miss a whole lot of traffic. If this happens, chances are someone might have to make physical contact with the commander to advise radio traffic is being missed.

You might be thinking you're safe if you never cover up your radio or turn the volume down, but this may not necessarily be the case. Scientifically speaking, you can't effectively listen to and comprehend multiple conversations simultaneously because listening is a conscious task, and your brain is capable of performing only one conscious task at a time.

You can quickly alternate back and forth between two conscious tasks and, thus, can monitor two conversations at the same time. Just understand you are at risk of missing some of the radio traffic and misinterpreting what you do hear.

An example that might demonstrate your vulnerability is to switch television channels back and forth in an attempt to watch two shows at the same time. This isn't exactly the same because you get the benefit of multiple senses (vision and hearing). If you want to make it more realistic, close your eyes and alternate between the television channels. You're going to miss clues and cues. The problem is that unless you have someone else there monitoring this experiment and telling you what you missed, you may not even know what you missed and this can, in turn, give a false sense of confidence that you didn't miss anything. After all, you have a pretty good idea of what's happening on both channels. Now, if you want to perform the ultimate experiment, alternate between two channels broadcasting the final 5 minutes of two fast-paced sporting events where the score is tied and both sports are of the variety where scoring is difficult and infrequent (e.g., hockey and soccer). Switch back and forth between the two. Go as fast or as slow as you want. Do it however your fancy strikes you. You are going to miss information. And if the announcer on one channel is hyping the listening for an impending score, your emotions may have you lock on to that channel a few extra seconds longer. That makes sense. You want to hear it. Just know the longer you stay locked on that channel, the more vulnerable you are to missing what's happening on the other channel. Thus, when you switch back to the alternative channel, you may have to stay there longer to figure out the status of that game because you missed so much. Remember—your eyes are closed, and you're only listening to each game, which means you can't look at the scoreboard. You have to wait to be told the score from the announcer.

Now, imagine that you switch from one channel over to the other, and you hear the announcer say, "That is unbelievable. I've never seen anything like that in my 20 years of covering this sport." You know you're going to stick around on that channel to see the replay of what you missed. Because the announcer is so charged with emotion, hearing that stressed voice charges you with emotion also, and it gets—and keeps—your attention.

You are also vulnerable if your brain mixes up the facts from one conversation with the other. And, you are vulnerable if you miss some portion of radio traffic and your brain decides to confabulate (make up) what it thinks you may have missed. There's a chance you could get it right based on what you are able to deduct. But, most unfortunately, there's also a chance you could get it horribly wrong.

Multiple radio channels

A similar outcome can occur if you try to monitor multiple radio channels. I often see departments who operate at incident scenes using multiple radio channels. Fundamentally, there's nothing wrong with that practice. In fact, for situational awareness, it's preferred if there is going to be a lot of radio traffic or if the incident operations can effectively be segmented by task to separate channels. The problem is not separate radio channels. The problem is one person attempting to monitor the radio traffic of multiple channels. Remember, listening is a conscious task, and humans can effectively listen to, and comprehend, only one stream of audible messages at a time.

Arguably, the more you are exposed to an environment of a multiple audio stimulus environment the better you could become at listening to multiple simultaneous conversations. This may be partially true. Emergency dispatchers may be a good example of this. It's not that they get any better at the physical act of listening; that remains a conscious task. Rather, they can hone the skill of filtering out the noise (noncritical pieces of information) and become better at listening for those keywords that are part of the most important radio messages they need to capture. It takes lots of practice and repetition to develop this skill set. Commanding a few incidents a month—heck, commanding a few incidents a day—is not likely to lead to this skill set. Remember, dispatchers listen to multiple conversations for hours on end every single shift, usually with headsets on to filter out all the extraneous noise. That level of repetition aids in their development of recognition of critical clues.

Best Practice #4

Be strategic when choosing the location for command.

For reasons founded in neuroscience and cognitive psychology, I recommend commanding from a vehicle. As mentioned earlier in this book, commanding from a quiet, remote location negates many of the barriers to situational awareness. However, if that is not possible, my alternative recommendation is to command from a physical location close enough to the incident's action that you can still maintain a visual fix on the activities of the incident, while far enough away to maintain a big-picture view and away from distractions and tasks that may entice you to become a "hands-on commander."

The closer to the action you are, the more vulnerable you are to having your situational awareness affected. In some ways this statement seems counterintuitive. One might argue that being up close and personal allows a commander to capture all the important clues and cues. To some extent, that argument is irrefutable.

The closer you are to the action, the more clues and cues you are exposed to, which is why I recommend every commander conduct a personal size-up before assuming a fixed command position (figure 19–3).

However, it's important to remember that being up close and personal to the action also puts you into an environment that is rich with stimulation (i.e., clues and cues to see and hear). This can quickly contribute to information (or perhaps I should say stimulation) overload and can quickly overwhelm you. Simply stated, there's just too much information to process.

You may recall from chapter 10 how stress changes the way your brain captures and processes information. Among the many consequences of stress is hypervigilance—putting your eyes and ears on high alert to detect potential harm. Your brain's limited capacity to process large quantities of information (and complex information) can be impaired under stress. As you get physically closer to the incident, it stands to reason you are exposed to more and more visual and audible stimulation

Fig. 19–3. A commander should think strategically about selecting a command location. Being out of the action and maintaining a big picture view of the incident are two essential requirements when picking the location for command.

(i.e., information). This can quickly lead to overload, and you may find yourself missing some of the most important clues and cues because your brain just can't process it all.

Gist versus detail

Another way to look at this is to apply the concept of *gist* versus *detail*. *Gist* is the essence, or the high-level, broad concept of something.

For example, if you have the gist of how your car operates, you understand it needs to have fuel. The fuel is ignited inside the engine and that makes the motor run, which makes the wheels turn. It's a fairly simple concept to grasp when you think about the gist of what propels your vehicle.

However, what happens when you get into the *detail*? You can quickly be overwhelmed with information that, under certain conditions, may be valuable, but often is simply more information (detail) than you need. For example, is it important for you to know the engine timing settings or the calibration settings for the brakes? Unless you're doing repairs on your car, the answer is no.

That's why you take your car to the repair shop, and you expect the mechanic to know (or have ready access to) the detailed information. After the mechanic assesses what is wrong with your vehicle, you want to know what's broken and how he or she intends to fix it (and probably what it will cost), but you don't want to know (and probably would not understand) the complex details of what went wrong. You want the gist, not the details. Let the mechanic (the expert on the front line) capture and process the details and then report back to you the gist.

The same concept can be applied to emergency scenes. The person in charge needs to capture and process a small quantity of the most important information. Remember, the capacity of the working memory is about five to seven pieces of information. Commanding from a vehicle or a remote location facilitates thinking and processing gist instead of detail. And while it may seem counter intuitive, it can be an asset to avoid getting overwhelmed by the massive amounts of information that can bombard you when you get too close to the action.

A creature of habit

Another, and no less compelling, reason to command from a vehicle or a remote location has to do with another cognitive response that occurs under stress. You may recall from an earlier discussion that under stress you are likely to revert back to behaviors that you're comfortable with or the behaviors that you have turned into habits. Let's apply this concept to an emergency scene where the commander gets too close to the action.

Let's say you have spent years developing your skills as a first responder. You are well trained and have hundreds, perhaps thousands, of emergency response experiences under your belt. When it comes to performing front-line street-level skills, you're on top of your game. You're good at it, comfortable with your knowledge, and competent in your abilities—all the skills a top-rate responder aspires to have.

If you're a front-line responder, the perfection of these skills is essential. They may even save your life. When highly developed, you can arrive, assess the situation, and take action.

However, if you are promoted into the role of commander, your role at the emergency scene is to serve as the big-picture coordinator of the event. You are no longer one of the street-level, front-line workers. Now your role is hands-off, serving as the overall coordinator of the entire incident.

A problem can occur when the big-picture, overall commander gets (physically) too close to the action. When this happens, the commander may feel compelled to engage in street-level, hands-on activities. In the commander's defense, the reason lies partially in the genetic response to stress, which compels us to do tasks that are comfortable or habit-based.

This may explain, in part, how commanders can be drawn off task and find their attention (or physical actions) trained on a front-line activity. The physical proximity of the commander to high-stress activities may actually evoke a strong (subconscious) urge to get involved in the front-line tasks. However, when this happens, the commander is no longer in command.

As established in chapter 9, the human brain (more specifically the working memory of the human brain) can only focus on one conscious task at a time. The instant a commander chooses to engage in a front-line, hands-on activity, the ability to see and comprehend the big picture is gone, and situational awareness (at least of the big picture) erodes. This *hands-on commander* may have wonderful situational awareness at the task level; however, and unfortunately, at the moment the commander chooses to be hands-on, no one is looking out for the well-being of the personnel operating on the larger, more complex overall scene (figure 19–4).

Fig. 19–4. We are creatures of habit. This is especially true under stress. If a commander has spent many years donning gear and SCBA in preparation for fighting a fire, he or she may (completely by habit) don gear and SCBA even when assuming the role of the incident commander. Established habits can be very hard to break.

When I press commanders for explanations as to why they perform hands-on tasks, I never get the response, "Because I was compelled to do it based on my hereditary instincts to revert back to doing what I'm comfortable with or what I have developed habits doing." They don't say that because they don't even realize this powerful hereditary-based, instinct-driven response is present. Rather I hear things like, "We are short-staffed, so I have to help pull the line," or "We needed a ladder raised to the second story window, and there was no one else to do it, so I had to."

I can't tell you, unequivocally, that a commander should *never* do anything hands-on. There are times when urgent, life-saving action may be required, and the commander is the only person who can take such action. And there may be times when staffing is so short that the commander must assist the crew at the task level. In either case, do what you need to do.

However, it's vitally important for you to understand when you perform hands-on activities, you are relinquishing the ability to effectively capture the clues and cues and make effective decisions at the command level.

Contributing to the danger of this situation are commanders who think they can justify their actions by saying things like, "I was only engaged in hands-on activities for a few minutes, then I was back to my commander role." Anytime you become disconnected from the big picture, there's a risk of degrading situational awareness. You're at risk of missing vital information while your eyes and ears are distracted from what they should be noticing.

You're also at risk of becoming confused because the clues and cues have changed and the incident you see when you return to your big-picture position may not fit the mental model you held prior to engaging in your hands-on activities. This may create confusion, even disorientation. It may take some time to sort things out and get reoriented. At a fast-moving incident, you're at risk of not being able to catch up after you fall behind because the incident changes so rapidly.

Location, location, location

Finally, the location of the person in charge determines the amount of distractions and interruptions they experience. Think if it this way. When you have a lot of work to do and you don't want to be interrupted, you go find some place quiet. There you can concentrate and get your work done with a minimal number of distractions and interruptions. The same logic applies to an emergency scene.

If a commander is physically located too close to the action—in the midst of loud noises and bright lights (i.e., lots of physical stimulus)—it stands to reason that attention is going to be drawn to those stimuli. As discussed in chapter 10, your genetic engineering makes you predisposed to be attracted to objects that are loud, bright, and moving—especially those objects or people that are moving toward you. Instinct tells your senses to go on alert and pay attention because those objects represent a danger to you.

Let me be clear on this. Those objects moving in your direction don't have to be adversaries or rabid dogs. They can be people you know well and they're coming to share valuable information with you. From the perspective of your genetic engineering, however, your body gives its attention to the person moving toward you, be he friend or foe.

Lights, cameras, action!

The closer you are (physically) to moving, bright, and loud stimuli, the more your attention is going to be drawn to them. Put yourself in a physical position at an emergency scene where people can have ready access to you, and you may find yourself quickly losing your situational awareness of the incident's most important visual and audible clues and cues as you find yourself watching and listening to loud, bright, and moving stimuli that are in close physical proximity to you. Some commanders (for reasons that may seem to be explainable only as ego-driven) like to don white bunker gear and white helmets or vests emblazoned with *Incident Commander* and stand in the middle of the action, almost as if to say, "Look at me! I'm in charge here! Look how important I am!" The best commander may be one that front-line workers (and the media) never see until the incident is under control. That doesn't mean the commander is remotely located and cannot see the action. It means the commander keeps a distance from the action and an eye on the big-picture incident.

The location of the commander can be a topic of much controversy. In fact, I have caused some heated discussions between supervisors and commanders by asking them, "Where should a commander be located? In a vehicle (what I call in the seat) or outside the vehicle (what I call in the street)?" It's clear from the responses that this is a polarizing topic. Each individual is passionate about where he or she thinks the commander should be located. Street commanders have all their reasons, and seat commanders also have all their reasons.

Among the discussions I've had on the topic of commander location, some have demonstrated a passion for commanding in the street. I ask these commanders if they could ever imagine a time when they would not command in the street. In other words, would they ever see it feasible to command from inside a vehicle? The frequent response is when the incident becomes complex and they have multiple companies working in the hazard zone, they believe they would retreat to the quiet calm of a vehicle where they could sort things out. Of course, this would only happen if the commander possessed strong situational awareness of how the environment was affecting information processing and decision making. Paradoxically speaking, it is the very environment that may impede the ability of commanders to realize they are in a bad physical location in the first place.

Many street commanders also admit, often in the same breath, that they get so overwhelmed with activities going on and the speed and the complexity of the event that they never retreat to a vehicle, noting they either forget to move or feel they don't have time to move. In those instances the commanders readily admitted to becoming overwhelmed very quickly and finding themselves either not knowing or forgetting some of the most important safety information—like the size and location of crews operating in the hazard zone. They get behind and never get caught up.

Equally, commanders who operate from a vehicle have their passionate reasons for doing so, most revolving around the need to control distractions and interruptions as well as being able to control environmental conditions (such as temperature and lighting). Seat commanders note it is easier to use checklists and worksheets; easier to have ready access to incident data such as preplans or hazardous materials information; easier to physically track crew sizes, locations, and movements; and easier to monitor radio traffic with headsets in an environment with little ambient noise.

I once asked a seat commander if he could ever imagine a time when he would get out of his vehicle and command in the street. His response was quick and direct: "If my vehicle ever caught on fire while I was commanding, I would get out of it and command in the street. Otherwise, no."

Regardless of the location from which you choose to command, know this: Commanding an incident requires a total commitment of your physical and mental resources to be oriented to the big picture. Keeping your distance from the physical, street-level activities can go a long way to controlling distractions and interruptions and improve your situational awareness.

Best Practice #5

Call a personal time-out.

When a sports team needs to assess how things are going or to regain their composure if things aren't going well, they call a time-out. As first responders, you should do the same thing (figure 19–5). Why? Because you're so busy paying attention to everything happening around you that you're not paying attention to yourself and what is happening to you. There's no mirror for you to look into to see if you have a look of calm or a look of terror on your face. There's no monitor connected to you that sounds alarms or displays flashing lights if you are losing your situational awareness.

The only way for you to know if you have strong situational awareness is to pause and take inventory of what you're doing and compare your situational awareness to the following benchmarks.

A personal time-out is a way for you to pause, if only for a few moments, to ask yourself (on a conscious level), "How is my situational awareness?" This is a critical task because it's not something you're likely to think about during the heat of a high-risk engagement.

Fig. 19–5. A situational awareness best practice involves occasionally taking a personal "time-out" and walking away from the stress and the action. This can allow the commander to gather his or her thoughts and conduct a personal situational awareness check up.

You can accomplish a personal time-out several ways. The first, and perhaps the most effective way, is to consciously benchmark the steps to forming situational awareness. (Note: Dr. Mica Endsley, whose definition for situational awareness is the one I discussed in chapter 4 and use throughout this book, suggests the progression of situational awareness be called *levels*, not *steps*.) Here is a refresher of the levels of situational awareness:

Level 1: Capture the clues and cues in your environment.

Level 2: Comprehend the meaning of the clues and cues used to interpret the current situation.

Level 3: Predict future events based on your assessment of the current situation and the likely progression of the incident.

Until these become part of your subconscious programming, I recommend having them written down on a worksheet, a checklist, or a wallet card. As you are commanding the incident, pause (take a personal time-out) and ask yourself (consciously), "Am I doing all three of these?" The prompt serves as a reminder to take time to assess if your situational awareness is strong, weak, eroding, or absent. It's like a mental tune-up of sorts. As much as you may not be able to control many of the events happening around you, you may be equally as vulnerable when it comes to controlling your personal situational awareness. There are lots of activities happening, both inside you and in your environment, that are trying to rob you of your situational awareness. Some you may be consciously aware of, many you may not.

If you have the levels committed to memory, you may only need to have a prompt written on your worksheet. Perhaps the prompt would be "Got Situational Awareness?" You are reminding yourself that in addition to all the actions and activities happening *outside* your being, there are things happening *inside* your mind that are equally as critical to the incident's safe conclusion, none of which are (arguably) more important than developing and maintaining your situational awareness.

If you developed strong situational awareness on your arrival at the emergency scene, you made some projections of the future events (i.e., Level 3 situational awareness). But how far into the future should you envision? There are many factors that form the answer to this question. For example, the speed of the incident has as much to do with how far out you look at anything. The faster the incident is moving, the shorter your view into the future. For example, at a fast-moving structure fire, rescue, or hostage situation, the prediction of future events may encompass a few minutes. At an event that may be moving slower, the prediction of future events might extend 10 minutes into the future. In some cases, such as extended disasters or wildland fires, the projection of the future could be days, weeks, or even months.

Generally, the focus of this text has been to look at the short-term events that cause immediate and catastrophic outcomes based on rapidly eroding situational awareness in dynamically changing conditions. Other texts provide information about how to do long-range planning, such as creating written incident action plans. I'm referring more to situations that can erode before you have a chance to even get out your pen and paper.

Let me pause here for a moment of clarification because I don't want readers to somehow interpret that last statement as I'm saying you shouldn't write things down. Quite the contrary, I think it's essential to use job aids like worksheets and checklists to help you manage your working (short-term) memory. Those forms should be at your fingertips at all times and should be designed for easy reference. I discuss this in chapter 12, so I won't belabor it here.

Okay, where was I? Oh yes, Level 3 situational awareness. As you arrive, you assess the situation and make predictions of future events. Let's say, for the sake of this example, you looked 10 minutes into the future and imagined the incident would look a certain way (based on your expectations of resource abilities and time to get the task completed).

Now the time has come. Like Marty McFly, you have arrived in the future. You're at the 10-minute mark, and you look at the incident's clues and cues. You must make an assessment if the incident looks like you thought it would when you developed your Level 3 situational awareness 10 minutes ago. If it does, then you know your incident is progressing in alignment with your mental model (your *movie* of the future) and that's a good sign your situational awareness is strong (or at least one component of it is).

However, if you look at the incident at the 10-minute mark and it doesn't look like it's supposed to (i.e., things look worse than anticipated), alarm bells should be going off in your head. Something's wrong.

And this is where the lesson of Level 3 situational awareness becomes critically important. If you never took the time when you first arrived to make a prediction of what the incident should look like at the 10-minute mark, then you have nothing to gauge your current situation against.

Stated another way, you have no way of knowing if you're making the progress you anticipated if you didn't set a goal and a deadline when you first arrived. That's a very dangerous place to be. It's the equivalent of getting lost in the woods, becoming scared, and starting to run wildly through the brush and trees in a desperate attempt to find your way out. What happens is you become even more hopelessly lost and disoriented. The same thing can happen at an emergency scene.

Let's say, hypothetically, that you did develop your Level 3 situational awareness and did make a prediction of future events for 10 minutes into the incident. Now you're at that point in time and things don't look good (i.e., the progress you thought would be made has not happened). At this point you have to make a decision. There are, essentially, only three decisions you can make.

Decision one: Extend the deadline by giving the crews additional time to accomplish the task. (You may also make modifications to the plan.)

Decision two: Assign additional personnel to the task, and extend the deadline. (This, too, may also come with a modification to the plan.)

Decision three: Withdrawal the resources from the situation, and take up a defensive posture.

So, which do you do? The obvious answer is that it depends on the situation. The less-than-obvious and more-accurate answer is that it depends on the ego and self-esteem of the commander. For more detail on this dangerous vulnerability, revisit chapter 13.

Best Practice #6

Use a command team.

The importance of sharing the mental workload at an emergency scene cannot be overstated. Commanding an incident that is high-risk, high-stress, and high-consequence and involves overseeing and coordinating multiple groups of responders is a task that requires the cognitive capacity of more than one person. As I read casualty reports from incidents where multiple crews of responders are operating in very challenging conditions and their actions are being coordinated by a commander working alone, I can't help but think that the commander was predestined for failure. If this book has established anything, it's that humans have vulnerabilities and shortcomings in cognitive capacity, and many of those issues are exacerbated by stress.

So, if we take 15 responders and put you along with them in a fast-paced situation, where clues and cues are changing rapidly, add lots of noise and radio traffic to be interpreted, add in all the side effects of stress on your ability to take in and comprehend the events that are happening around you, it's easy to see how some incidents could be set up for catastrophe. Well-intended commanders, regardless of training and experience, can get overwhelmed and overloaded if they are working solo.

The solution is to build a command team as soon as feasible. Providing the commander with an aide or advisor 30 minutes into an incident isn't going to do much to reduce catastrophic outcomes that often occur in the first 10 to 15 minutes of the event. It's very unfortunate that many departments operate with single command officers at all but their largest (multiple alarm) incidents.

The Chauffeur

At one time (perhaps when we lived in a more robust economy) command officers had drivers/aides. As the economy tightened, these positions were among the first to be cut, citing them as unnecessary luxuries to have the commander chauffeured to the scene of an emergency. From the perspective of situational awareness, this shortsighted, budget-driven philosophy could not be further from the truth (figure 19–6). Here's why.

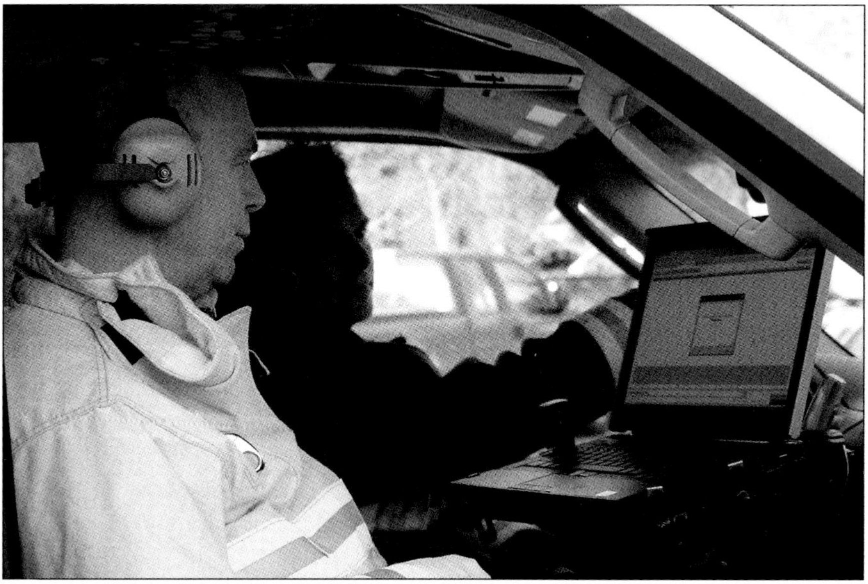

Fig. 19–6. The chauffeur can be a valuable asset for a commander, as this relieves the commander from the cognitive workload needed to safely drive the vehicle to the emergency scene. This can improve the commander's ability to process and comprehend critical clues and cues over the radio and look up valuable preplan information while en route to the emergency.

Remember from earlier discussions that your brain can perform only one conscious task at a time, and the conscious brain has a limited capacity when it comes to how much information it can process at one time (about five to seven pieces of information). You may also recall that your conscious brain can switch back and forth among several tasks, giving the appearance of multitasking, when, in fact, you're single-tasking in rapid sequence. The subconscious brain (unlike the conscious brain) can perform multiple tasks simultaneously, as long as those tasks have been stored in the subconscious brain through repetition and memorization.

With a review under your belt, let's look at what can happen when a commander gets behind the wheel and drives to an emergency scene. Let's first take the obvious act of driving the vehicle. Wait—you can do that subconsciously, right? Driving is something you have done over and over again, and most experienced drivers operate their vehicles using their subconscious brains. If you were thinking that already, give yourself a big gold star. You're absolutely correct. Assuming you're a highly trained and highly experienced driver, you can operate a motor vehicle using mostly subconsciously stored cognitive tasks ("What am I supposed to be doing?") and muscle memory tasks ("How am I supposed to do this?").

However, before you affix that gold star to your wall of fame, you may want to stop and contemplate something. You can operate a vehicle using subconscious, scripted programming only as long as everything that's happening while you're driving is routine and expected—traffic conditions are normal, and other drivers are behaving in the highly predictable ways you have stored in your memory based on years of observing them while you drive. See where this is going?

What happens when you turn on your emergency lights and siren and respond to an incident? Do traffic conditions change as a result of your approach, and perhaps the use of traffic signal pre-emption equipment? Do the highly predictable behaviors of fellow drivers become highly unpredictable when they see lights and hear sirens? If you're an experienced emergency vehicle operator, you know firsthand that the act of driving changes completely when you are operating an emergency vehicle in response mode. Thus, the act of driving no longer is a subconscious, automatic scripted routine. It's very much a conscious, unscripted task that requires a great deal of cognitive resources to develop and maintain your situational awareness simply on the act of responding.

Unfortunately, that's not the only thing going on in that high-stress, high-consequence emergency response environment. You are also being bombarded with clues and cues about the incident to which you're responding. Information is being relayed from dispatch about the call, and you're trying to keep track of which units are responding and from what direction they are coming (both to aid in tactical deployment and to avoid meeting up unexpectedly in an intersection). You're trying to develop a mental picture of the incident scene in preparation for assignments on arrival. All of this requires conscious (working) memory. The only problem is, this environment makes you prone for information overload and something may get lost. You may lose important pieces of information about the incident's clues and cues, or you may lose important information you need for driving. Perhaps now it's easier to understand how one of the leading causes of first-responder casualty events (second only to heart attacks and strokes) are vehicle accidents.

In theory, if commanders driving themselves to emergency scenes are vulnerable to cognitive overload, then so are apparatus operators who are trying to listen to and comprehend radio traffic while operating very heavy vehicles that behave very differently than the passenger vehicles they drive day in and day out.

Commanders who are relieved of the act of driving are at a distinct advantage when it comes to developing and maintaining strong situational awareness in the critical first few moments of an emergency, in which most of the clues and cues are coming in the form of verbal messages that must be processed by the brain into visual images. A commander unencumbered by the act of driving has the ability to listen to and comprehend much more information. When I had a driver, I put on my headset and used my worksheet to keep track of things while we responded. Sometimes when dispatch offered complex details, I closed my eyes to form the visual image of what they were saying. I didn't want to be looking at the roadway and traffic; it was distracting. That was but one advantage where my driver helped me develop strong situational awareness.

Captains and First Officers

Let's look at this from another angle. There's a reason commercial airlines have a captain and a first officer. Each has specific roles. And while there are two pilots, only one flies the plane at any given time. In fact, in aviation they have acronyms for this. The *PF* is the pilot flying the plane, the *PNF* is the pilot not flying the plane, and their roles and duties are clearly delineated. These roles are most important to maintain during the times when the pilots are performing high-risk, high-consequence activities (e.g., take-offs and landings). Both pilots are busy with distinctly different tasks. One pilot flies the plane, but they share the cognitive workload.

Two people in the command vehicle while responding frees up the commander's cognitive capacity to focus on the radio traffic, keep track of responding units, and form mental models. When on scene, they become a command team—combining the mental capacity of two people to capture and keep track of clues and cues in the dynamically changing environment. After the personal size-up, the commander establishes a fixed location for the command post while the aide/advisor remains mobile, gathering information about the incident. The commander stays focused on watching the incident's changing conditions while the aide/advisor tracks information on the worksheets and checklists. The command aide/advisor also listens attentively to the radio so critical radio traffic is not missed. All of these support tasks free up cognitive capacity for the commander to think. With the improved capacity, the commander is far more likely to capture the right clues and cues (Level 1 situational awareness), process those clues and cues into meaning (Level 2 situational awareness), and make accurate predictions of future events—mental modeling (Level 3 situational awareness). And there you have it: a commander with strong situational awareness.

Best Practice #7

Control your distractions and interruptions.

You may recall there is a difference between distractions and interruptions. Distractions draw your attention away unexpectedly (e.g., a loud noise, a scream, an explosion, a flash of light, and so on). Interruptions draw your attention away with purposeful intent (e.g., radio traffic calling you, receiving a cell phone call, someone approaching you to have a conversation, and so on).

This best practice is tied closely to many of the others and perhaps is a direct by-product of the other best practices. For example, commanding from a vehicle or a remote location is a way of controlling distractions and interruptions. Likewise, so is avoiding being a high-profile target (e.g., wearing white turnout gear and standing in the front yard).

Distractions and interruptions are significant enough barriers to situational awareness that they deserve their own best-practice recommendation. It's absolutely essential that you understand what can happen as a result of distractions and interruptions.

You are far more likely to miss critical clues and cues when you are distracted or interrupted. And you are far more likely to forget the clues and cues you have already processed, especially if the distraction or interruption elicits an emotional reaction from you. You are also very likely to forget where you were on your original task when the distraction or interruption occurred. Finally, it can take you a while, cognitively, to clear your head of something that distracts or interrupts you, contributing further to an eroding situational awareness and the recovery.

Distractions and Interruptions

How do you control your distractions and interruptions? No doubt it can be difficult while working in a fast-paced, stimulus-rich environment. One of the best ways is to think about how you have been distracted and interrupted in the past. What drew your attention away from the most important clues and cues you should have been paying attention to? After you identify the barriers causing the distractions and interruptions, set out to develop a strategy to prevent them or to quickly overcome their ill effects.

For example, if you're a street commander and you find yourself being distracted by the loud noises, bright lights, and moving objects that are happening around you, then consider removing yourself from that environment. The problem is realizing how much you are actually being impacted by the distractions. As another example, if you are commanding an incident and someone comes up to you to share information, they're interrupting your cognitive processes. The problem is, you don't really know if what that person is about to tell you is important or just noise. And you won't be able to make this assessment until you actually hear and process it. But by that time, it's too late. Your conscious capacity to pay attention to the incident was lost because the act of listening is a conscious act. As long as you intently listen to what someone else is saying (in person, by radio, or by phone), it's impossible for you to coherently comprehend anything else being said. Simply stated, you can't comprehend two simultaneous conversations without putting yourself at risk of missing something because your conscious brain alternates back and forth between the two conversations.

A gentleman once shared story with me about a Sunday afternoon while he was intently watching a football game on television. During a very tense portion of the game, his wife decided to engage in a discussion on a very important topic—or at least it was very important to *her*. He didn't want to be rude and tell her he was busy and didn't want to talk to her. Equally, he didn't want to turn off the game. So he made an ill-fated decision to multitask the act of listening—which he thought he could do but found out through this experience he couldn't.

As his wife spoke he strained to listen to her message intermixed with the very excited and emotional commentary from the announcer on the television. He noted he provided his wife with what he thought were appropriate feedback cues, lots of head nodding and smiles. But, while he gave the outward appearance of listening to her, the visual cortex of his brain, processing those audible messages into images, was fixed, almost entirely, on the much more stressful and interesting message from the announcer on the television.

He felt accomplished at his successful demonstration of audible multitasking. About a week later, something important his wife had said in the conversation was brought up again. He had no recall of what she said. Sparing you the harsh details, this caused him some consequences.

When operating in a high-stress, high-consequence environment, it is entirely plausible that you may fail to realize the impact distractions and interruptions are having on your cognitive processing capabilities. You may miss and/or forget clues and cues and never even know it. Why? Because you cannot be aware of what you missed if you never knew about it in the first place. Of course, what you missed may come back to you in a painful after-action assessment where investigators (and perhaps lawyers) may gather video and audio from the incident and use it to demonstrate the seemingly incompetent behavior of the commander who missed critical clues and cues.

If you could retrospectively look back on previous near-miss events, you might be able to identify times where distractions and interruptions affected your situational awareness.

For additional examples, look to the wealth of lessons contained in the National Near-Miss Reporting System (www.firefighternearmiss.com) or the casualty investigation reports from the Center for Disease Control (CDC) division, National Institute of Occupational Safety & Health (NIOSH) (www.cdc.gov/niosh/fire/). When you read these reports, be open minded. If it could happen to them, it could happen to us.

Best Practice #8

Develop and maintain a strong command presence.

Command presence is the ability of the person in charge to, well, be in charge. This requires a person who is well trained, competent, confident, disciplined, and able to lead and make decisions. Equally, it requires a person who can stand up and do the right thing when it comes to ensuring the safety of personnel—even when doing the right thing is terribly unpopular. Unfortunately, there are some commanders who would rather be popular than effective.

The desire to be well-liked and respected among peers and subordinates is a powerful emotion and should not be discounted when evaluating the effectiveness of commanders. Strong commanders need balanced self-confidence and strong self-esteem. They cannot let ego cloud good judgment. Competent commanders are able to accept constructive criticism and suggestions for improvements. Command presence means refraining from becoming angry or defensive if the incident isn't progressing as planned and being mature when blame is pointed in your direction. Commanders have the same qualities that are in all of us. They are fallible human beings with specialized training, specialized experience, specialized abilities, and some special limitations.

Emotional Control

Having a strong command presence means having the ability to hold your emotions in check. Note that I'm not saying commanders shouldn't use emotions to help in their decision making. In fact, research has proven that with the absence of emotions, decision making is flawed. What I'm suggesting is controlling emotions like anger and frustration. There are times when situations don't go well, and people under your command make mistakes. If you become angry and frustrated when this happens, it may draw your attention away from the important incident objectives and command priorities.

When a commander gives a tactical assignment to a crew, the commander should form two expectations about the performance of that task. First is an expectation of how the incident benefits from the completion of the task. Second is how long it should take for the task to be completed.

For example, assume a commander gives a crew a task and expects that the completion of the task will effectively neutralize the incident. Further, let's assume the commander expects the task to be done in 3 minutes. However, at the three-minute mark the task is not completed. The commander begins to show a little concern. At the four-minute mark the task is still not completed, and the incident is getting progressively worse. The same conditions (worsening) are present at the five-minute mark. At this point, the commander should become concerned for the well-being of the crew because their objective is not being completed to the commander's expectations.

However, what may happen is the commander doesn't become concerned. The commander, instead, becomes angry and frustrated, struggling to understand what in the world the crew is doing that is taking them so long to get the task done. If this causes the commander's stress level to rise, it may also cause the commander's attention to narrow down to the problematic crew, contemplating where the crew is, what they are doing, and why they haven't completed their assigned task on time. If the situation doesn't improve in a hurry, the commander may become so frustrated and preoccupied with the crew's location and progress that other important tasks and the accountability of other personnel may become secondary. In a worst-case scenario, a frustrated commander may feel compelled to become hands-on and perform the task alone or at least join up with the crew to provide proper oversight to the task completion. As noted previously, such a decision effectively takes the commander out of the command role, leaving the incident with no one in charge. It's tough to have command situational awareness when there is no one commanding. Emotional control is an essential trait of a successful, focused commander.

Strong command presence also means having the ability to direct the operations of personnel—to set a strategy and communicate it effectively so all personnel are on the same page and moving the incident toward a commonly understood goal. To accomplish this, the commander must be able to control the assignments and movement of personnel (i.e., the crews perform as directed by command). Freelance operations are not permitted.

This also entails a commander's willingness to give directions and hold personnel accountable for task completion.

I once had a commander tell me that when he arrived on the scene of an incident he didn't have to do much to coordinate the activities of crews. He boasted that his crews were so well trained and their assignments were so pre-scripted that the crews essentially knew what to do without having command meddling into the operations. Were they indeed that good, or did the commander lack the abilities to effectively lead? The conditions of every incident vary and an essential component to overall incident success comes from having someone who monitors the conditions and directs personnel based on a dynamic risk assessment, this commander was either overconfident in the abilities of his crews, or he lacked the ability to effectively lead. Either way, his command presence was substandard.

Even well-trained and highly experienced crews need someone coordinating their activities if for no other reason than multiple crews operating at an emergency scene are rarely in a physical position to see the outcomes of their efforts, much less the efforts of all the personnel operating at the scene.

At the very minimum, it's the commander's role to ensure a strategy is communicated, tactics are coordinated, accountability is maintained, and the radio traffic of all at-risk personnel is monitored. Running a dynamically changing emergency scene cannot be done on autopilot.

Command presence also means keeping track of the passage of time. This can be difficult when you're under stress. Heck, this can be difficult when you're having fun. Whenever your attention is preoccupied (by whatever), you are at risk of losing track of time. In scientific parlance it's called *temporal distortion*. It simply means time gets away from you. Five minutes may pass like 20, or an hour may pass and it seems like you've barely been on the scene for half that time.

Command presence requires keeping track of time. This serves many purposes. Time—accurate time—allows the commander to anticipate crew resource needs. Time benchmarks serve as good indicators for when crews need to be rotated to rehab or for a bottle change. Time benchmarks can also be indicators of how much stress the structural components are under, depending on how long they have been exposed to the fire. Time benchmarks also indicate how much weight is being added to the building if large volumes of water are being applied.

The commander who can keep track of the passage of time is better at anticipating outcomes. Temporal distortion, however, is not limited to the commander. All responders operating at an incident scene can suffer from temporal distortion. If they know how much time is passing, they may be in a better position to recognize their own fatigue or put into perspective how the incident is improving (or degrading).

> One of the best ways to keep track of the passage of time is to have the dispatcher announce *elapsed time notifications* (ETNs for short). Ideally, ETNs are announced over radio channels heard by all personnel operating at the incident scene.

How often ETNs are transmitted is a matter of preference. I've seen them announced at 10, 15, and 20-minute increments. My preference is every 10 minutes for ETNs and then a personnel accountability report (PAR) conducted every 20 minutes.

Best Practice # 9

Develop and train to scripted procedures but build resiliency into training programs.

This best practice may hold the distinction of being both a great asset and a great liability. Generally, public safety agencies perform better, are more coordinated, and have better results when they develop and use standard operating procedures (SOPs) or standard operating guidelines (SOGs). These documents ensure personnel are trained to a common set of expectations and will, generally, perform all tasks the same way. This predictability aids in the development of Level 3 situational awareness because it helps personnel be able to predict the future based on standardized procedures. This includes predicting (anticipating) the actions of fellow crewmembers that perform their activities to their scripted SOPs or SOGs. What could possibly be bad about that, you might ask. Isn't consistency and predictability of personnel following SOPs and SOGs a good thing? In general, yes, but not always. Let me explain.

The Downside of Standardized Procedures

Some personnel take SOPs and SOGs as literal instruction manuals for performance. They train to the procedure or guideline with the expectation of performing exactly as scripted all the time, without deviation. That may work well for them in most cases; however, emergency incidents are much too dynamic to develop scripted procedures that cover *every* situation. This is why I prefer the term *guidelines* instead of *procedures*. But what's in a name? It's more about the mind-set of the administration and how personnel are trained that will determine success or failure.

If the administration holds personnel steadfastly accountable to following the standard operating procedures with no margin for latitude, then the organization develops a cadre of script-following responders. This is especially true if there have been instances when the procedures were not followed and resulted in some negative consequence (perhaps disciplinary action) for the person(s) who decided the standard did not appropriately fit the situation. You have to beat only a few people into SOP submission before everyone learns to follow the scripts without variation. This can be a setup for failure.

Training for failure

Likewise, department trainers can teach response personnel to perform based purely on scripts. This can be viewed as an effective way to get all personnel in lock-step formation. This can, however, be very problematic as well. Personnel who are taught rote performance procedures by training officers may be set up to fail when faced with a novel event that doesn't fit the standard.

It would be better if trainers taught personnel to perform based on the SOPs or SOGs and then gave the responders novel situations with little or no guidance on how to solve the problem. Initially, trainers are likely to see personnel attempt to apply the SOPs and SOGs to solve the problem. When it doesn't work, the crew experiences a failure. Trainers are well advised to resist the temptation to correct the behavior. Instead, allow the incident to proceed to failure (with regard for safety). Unless the personnel are in danger of getting hurt, failure is an important component of learning. Through the lessons of failure, personnel build up a repertoire of experiences of actions that do not work. These lessons can be as important as knowing what does work. Better to learn what does not work in a safe, training environment than to learn those lessons for the first time at an emergency scene where the crew is going to get only one chance to do it right.

Creating resilient problem solvers

Training officers who take pride in their work may strive to have personnel perform routines over and over again until they can perform them flawlessly, and the flawless performance is seen as successful performance by the training staff. Initially, this is a good strategy. But if it ends there, the trainers have failed to teach the responders to be resilient problem solvers. Face it, just about anyone can be taught to follow a policy, script, or recipe. The most valued team members are those who can face novel situations and devise unique solutions on the fly. This does not come without practice, and that practice often involves—you guessed it—failure!

Where personnel can fail in training without fear of retribution, and trainers can develop a tolerance for failure, the organization is well on its way to improving operational safety. Failure in training may also improve responder safety. Failure in safely controlled ways develops improvisation skills and can improve resiliency when faced with novel situations. The creative problem solver is far more likely to have a good outcome.

The lessons from failure are important

How does failure tie to situational awareness? In chapter 3 I discussed how the brain captures and catalogs your experiences and then searches for solutions to problems by going through its catalogs (referred to earlier as recognition-primed decision making). Among the experiences the brain stores is not only how to do something, but also *how not to* do something. In other words, the brain stores and catalogs failures as well as successes. In fact, some science might suggest the experiences of failure are seated deeper into memory if the failure evoked a strong emotional response (anger, frustration, embarrassment, fear, and so on).

Best Practice #10

Accelerate your command knowledge and expertise.

Chapter 3 discusses how recognition-primed decisions come from an accumulation of all your life's experiences, training, emergency responses, videos, case studies, *MacGyver* episodes you watched in the 1980s—everything.

It can take a long time to accumulate enough emergency response experiences to prime your recognition. This is especially true of the types of incidents you may not see very often, such as structure fires, hazardous materials incidents, or hostage situations. Ironically, the types of incidents you are likely to see less of are the same types of incidents that can put your safety most in jeopardy. They are sometimes referred to as *high-risk, low-frequency events*. If high-risk events occur infrequently, then how can responders possibly obtain the experience needed to make quality recognition-primed decisions?

Surprisingly, the answer lies in your ability to trick your brain. Both psychologists and movie makers have known for a long time that the human brain can have a very difficult time distinguishing facts from vividly imagined fictional events. This can have unpleasant consequences if you're watching a scary movie or if you have developed a phobia based on your imagination. But it can also serve as one of the best assets in your arsenal to improve your safety at high-risk, low-frequency events.

The obvious way to preload your experience is through training (figure 19–7). The more you train in preparation for incidents, the more likely you are to be able to tap into those experiences when you are under stress and your brain searches for solutions to the problems you are dealing with. There's an additional, equally powerful, and far less expensive way to accelerate your collection of experiences. If you read case studies and line-of-duty death reports and watch videos of incidents, you can accelerate your learning. But there's a catch.

Fig. 19–7. A first responder can accelerate his or her learning by using high-fidelity simulators and practicing decision making in high-stress environments.

Vivid Imagination Is the Trick

When you're learning from incident reports, books, magazine articles, or videos, you cannot simply passively read these documents or watch the videos and expect to develop your expertise. The secret is to trick your brain into thinking the experience is real and not imagined. This is best accomplished by attaching emotions to the learning experience.

For example, if you watch a poorly produced (B-rated) scary movie and it's so bad that you are not scared at all, you're probably not going to remember much about it. However, if you watch a well-produced scary movie and it frightens you so much that you jump in your seat and your heart races, you are much more likely to remember it. I can vividly remember the houseboat scene from the movie *Cape Fear*, even though it's been many years since I've seen the movie. This is because the entire plot and the scary scenes were so believable that it was burned into my memory. The lead character was psychotic enough that you feared him, but not so far over the top that you laughed at him. And the things he did throughout the movie were realistic. The movie evoked an emotional response and, hence, I remember it.

When you read case studies and line-of-duty death reports and watch incident videos, you also need to vividly imagine being there—immerse yourself in the events and trigger emotional responses. Your brain stores those experiences in your mental slide tray as if you actually experienced them. Then, when you are under stress and your brain is searching for the solution, the reports and videos become valuable triggers to prime your decision making.

Likewise, when you train, the more emotion you can build into the training session, the deeper the lessons seat into your memory. What emotions do you have to evoke? The answer is easy—any. It doesn't matter whether you're happy, sad, angry, or frustrated. Any emotion will do, and research has not demonstrated any one emotion to be more effective than the other.

Are you kidding me?

Is there a place for humor in learning? You bet there is, and I use it—a lot. Appropriate humor that gets participants laughing aids learning in several ways. Laughing is an emotional response, and it triggers the release of neurotransmitters that not only make you feel euphoric, but also open the brain to be more receptive to storing experiences for long-term recall.

Additionally, stressed brains don't learn the same as the calm and relaxed brain. This lesson applies more to the ill effects of chronic stress on the body. But it stands to reason that if you have students who are experiencing chronic stress, the more you can get them relaxed about the lesson being taught, the more likely they are to learn and recall what you are teaching them.

I like using self-deprecating humor when I teach because it's always safer to make fun of myself versus others I may not know very well. However, I have developed a pretty good sense of who in my audiences are the jokesters, and when they are willing to throw humorous barbs at me, I enjoy bantering them back. Again, if you use humor, ensure it's always in good taste and never crossing the line of being discriminatory or harassing. That's not funny, and if you create an environment where students feel threatened, you harm their learning, not aid it, because they shut down to what you're teaching.

Repetition, Repetition, Repetition

Another way to accelerate your experience and expertise is through repetition. The more you practice, the better you remember. This is especially true when you memorize. There are two forms of memorization—cognitive memorization and muscle memorization.

Cognitive memorization is memorizing information such as people's names, phone numbers, e-mail addresses, multiplication tables, the alphabet, and so on. When you memorize and practice something over and over again, it may store deep into your memory and be retrievable for life.

Using memory aids can be helpful as well (such as the trick that memory experts share for how to remember people's names). For example, if you meet someone name Matt Marshal, you might think of a door mat where a federal marshal lives and the mat has a five-point-star-shaped badge on it. You envision the image when you look at him and you remember his name—Matt Marshal.

I recall in elementary school I had to learn the names of the five Great Lakes. Without revealing my age (suffice it to say that was a long time ago), to this day I still remember them. Can you? Huron, Ontario, Michigan, Erie, and Superior. I am able to recall this so readily because I was taught by a creative teacher that the first letter of each lake, when combined, forms the word *HOMES*. Do memory aids work? They do for me.

The other form of memorization is muscle memorization. If you physically perform a task over and over again, you can perform that task with no conscious thought whatsoever. The ability to perform the memorized physical task resides in the subconscious brain. This can be both good and bad. Under stress, your performance may be based on scripted routines, and if your training was flawed, your automatic performance may also be flawed. For example, when responders are training on interior fireground operations (be that at a burn building, an acquired structure, or using the fire station) they should be completing a 360-degree size-up at the start of every evolution. This redundancy builds muscle memory that becomes the automatic scripted performance under stress. Failing to build this muscle memory can result in responders arriving at real working structure fire incident and not completing a 360-degree size-up. Why? Because they didn't include it in training.

The Blank Slate

What happens if you have no memorized response, either cognitive or muscular, because you are facing a novel event? When this happens, you may be vulnerable to make mistakes, and it may take you much longer to accomplish the task. Also, be aware that novel tasks must be performed with the conscious brain, which means you must give your full attention to that task (remember, the conscious brain cannot multitask).

Simulations

You can also use simulations to preload experiences into your brain, but you have to be careful here. There's no shortage of simulation software packages on the market these days, and there is virtually no research being conducted to validate the ability of these simulators to create and store meaningful, relevant, fully recallable experiences. As a cognitive researcher and an experienced fireground commander, this troubles me because first responder organizations are buying software programs to train personnel on how to make critical decisions.

Simulators with high-fidelity, realistic graphics and relevant audio are the most likely to aid in learning. A simulator with slow, jerky-moving cartoon characters and animated graphics of incidents where the movements and progression of the incident are not realistic (and they're out there . . . I've seen them first hand) are probably better at teaching firefighters how to use a game controller than they are at building any level of command decision-making expertise.

It's a simple concept—the more realistic the simulation, the better the learning. After reading this book, if you are convinced that the lessons and issues I discuss are real and can happen at your incident scenes, then compare your simulations to the lessons in this book. If your simulations create realistic barriers to situational awareness that your commanders must adapt to and overcome, then I would give it my seal of approval.

The downside of low-fidelity, unrealistic simulators is the people training on them may be lulled into thinking they're developing valuable command experience because they're spending a lot of time with the simulator. In reality, they may be learning very little that's going to help them when they get into real-world scenarios.

Golfing on a simulator

I am a golf fanatic and have played since I was a small child. When I was a teenager, there was a simulated golf game in the arcade at the mall. I liked playing it because it was fun. There was a large white control ball in the center of the console, and when you rolled your hand over it in a rapid forward stroke, the golf ball would be launched down the fairway. If you rolled the control ball to the left, the golf ball would draw (go left, if you're a right-handed golfer). If you rolled the control ball to the right, the golf ball would fade (go to the right). The same basic premise guided chip shots and putting. I got really good at it, often shooting under par. But it didn't do a darn thing to improve my real golf game because there was nothing at all realistic about the game. As technology advanced, along came Wii and a golf game where you hold the controller in your hand and swing it just like a real golf club. Hmmm, I don't seem to be as good at this golf game because it reflects the subtle nuances of my real golf swing (which is to say, I'm not very good at golfing and the Wii knows it). If the Wii has done anything to help my real golf game, it has taught me to slow down my swing because the harder you swing the Wii console, the more erratic the outcome, just like when you're playing real golf and just like I have experienced in my golf game throughout my life.

Get a Mentor

The final advice I have to help you accelerate the development of your expertise is to have a mentor (or coach if you prefer)—someone to work with you as you develop your command skills. This is critically important because new commanders are likely to make mistakes. When this happens it may result in a near-miss or casualty event. A mentor might be able to help prevent such an undesirable outcome and, if for that reason alone, it's essential to have a trusted senior advisor.

However, there's an even more compelling reason. Inexperienced supervisors, in the process of learning how to be effective incident commanders, may make mistakes, have no ill effects, and therefore fail to realize that what they did was flawed.

Absent someone to point out their shortcomings, these mistakes may be repeated over and over again without consequence. Repeated enough, they become the commander's performance standard. As this happens over time, the inexperienced commander now becomes an experienced commander with flawed command skills. Absent a consequential outcome when the commander performs incompetently, he or she doesn't know better. This, in turn, gives the commander the confidence to believe his or her command skill set is strong when, in fact, it may be seriously flawed.

The longer this situation is left unchecked, the less likely the commander will be able to see the shortcoming. In fact, with the passage of enough time and the development of enough confidence, the commander may actually become arrogant about how good a commander he or she has become. This can develop into a dangerous mind-set.

The Recipe for Catastrophe

I regret having waited so long in this book to offer you the recipe for catastrophe that I developed based on my evaluation of near-miss and casualty reports. It goes like this:

> Incompetent behavior . . .
>
> Without a consequence (no casualty) . . .
>
> Leads to overconfidence . . .
>
> And contributes to complacency and arrogance . . .
>
> Precursors to catastrophe

A mentorship program can help alleviate this problem because it corrects the incompetent behavior before it has an opportunity to develop into a habit.

Of course, this assumes the mentor does not have a catalog of incompetent behaviors that are being passed down to the protégé, which is entirely possible. In cases where this has happened, the entire command team of an organization may have blind incompetence because no one sees the errors of their ways. When an organization with such deep-rooted incompetence suffers a catastrophic event, they sometimes justify their actions by saying things like, “That’s the way we’ve always done it here, and we’ve never had something go wrong.” Some may even try to diminish their role in the casualty by believing it was a combination of fluke events that was the culprit, not their incompetence. Here the ugly face of arrogance manifests as denial of all culpability.

CHAPTER 20

TWENTY-FIVE-POINT COMMAND HEALTH CHECKUP

As I evaluated thousands of near-miss and casualty reports I started to see a pattern develop. There were common themes among many of the incidents. This compelled me to create a list of the most common errors I was observing that could be corrected. The beauty of this list is many of the items require very little, if any, financial resources other than what it might cost to develop procedures and train your personnel to implement them. But the payoff can be huge.

As you read down the checklist, pause and reflect on your own organization's performance at emergency scenes. Be honest. You're the only one keeping score. If there are duties or responsibilities you are not fulfilling (or you're not sure if you're doing them), then set out to fix them. This *health checkup* is not an instrument designed to evaluate the appropriateness of your strategy and tactics or, for that matter, the effectiveness of your decisions. In fact, having a good outcome at an incident does not always correlate to the quality of your decision-making processes or the presence of well developed and maintained situational awareness.

Twenty-Five-Point Situational Awareness Checkup

1. The incident operated with one clearly defined commander at all times.
2. Command was passed only when necessary, and only after a briefing.
3. Where possible, the commander (or a designee) completed a 360-degree size-up of the incident.
4. The commander refrained from performing hands-on, street-level tactical duties.
5. The commander maintained a physical distance from the incident sufficient to ensure a big-picture view.
6. The commander developed and communicated an appropriate strategy to all personnel.
7. The commander gave assignments to personnel and coordinated the incident's activities.
8. The commander maintained accountability of personnel, their location, crew size, their actions, and their progress.
9. Incident communications were clear, concise, and well understood.
10. If additional resources were needed, they were called using a preestablished system of notification (e.g., a mutual aid box alarm system).
11. The tactics were properly matched to the size and complexity of the incident.
12. The commander developed and maintained a broad perspective of the overall incident.
13. The commander was able to think ahead of the incident and effectively predict future events.
14. Incident radio traffic was disciplined and manageable.

15. Someone was assigned to monitor every radio channel or talk group in use.
16. Every communication from personnel operating in a high-risk environment was heard the first time transmitted.
17. The incident had adequate staffing to carry out the strategy safely and effectively.
18. As the incident conditions changed, so did the strategy and tactics.
19. Progress reports were clear, concise, accurate, timely, and informative.
20. The incident was not hampered by conflicting orders or conflicting tactics.
21. Flaws in organizational culture did not adversely influence the strategy, tactics, or operations.
22. An incident safety officer was assigned and performed his or her duties appropriately.
23. There was equipment, apparatus, and resources at the incident scene to safely and effectively accomplish the mission.
24. Personnel operating at the incident were adequately trained to perform their assignments.
25. Proper standard operating procedures and/or standard operating guidelines were established, implemented, communicated, and followed.

Use this checklist to develop a survey that is completed after each incident. Ideally, *everyone* who operated at the incident scene should answer the questions. If the survey is completed only by the commander, you are at risk of missing an important perspective of others who were operating on the emergency scene and may have been unaware of what tasks were and were not completed. Only by capturing the broad perspective of all personnel are you completely sure all of your bases were covered.

It's entirely possible that some of the personnel may not know if some of the tasks were completed because they were not in a position to see or hear it was done. Thus, the survey response options should be:

___Yes

___ No

___ Unsure

___ I was not in a position to evaluate this item

Finally, if the checklist reveals shortcomings, don't look to fix blame. Look to fix the problem. This is an incident evaluation tool, not a personnel evaluation tool. Strive to use it for its intended purpose—making incremental improvements in the safety of your operations.

CHAPTER 21

EPILOGUE

Congratulations, you are at the end of the book. Well, actually, you're at the beginning. I attempted to write this book in a way that you would find both informative and mildly entertaining, although I would be terribly disappointed in myself if I didn't find a way to convey the content of this book as a *call to action*.

I would like to pause here and share with you how delivering this message to tens of thousands of public safety responders over the years has left me feeling both euphorically happy and dismally disappointed.

My Euphoria

There are probably few compliments that have had more impact on me than when veteran public safety officials come up to me after programs and say things like:

> I can't believe that I have never been taught this stuff! I've been a first responder for more than 30 years, and no one has ever explained my vulnerabilities to me like this. I don't know how I haven't had anyone killed yet. This has been a wake-up call. I've

taken pages and pages of notes of things we need to go back and change.

I can relate very well to what these veterans are saying. I was feeling the same way as I was learning this stuff for the first time after serving for 25 years. In fact, as I continue to learn the valuable lessons of neuroscience and apply them to public safety, I'm continually amazed at how positive the feedback is from responders and how much they appreciate learning about the science that can improve their safety.

My Disappointment

I've been blessed for the opportunity to share my message with so many first responders throughout the world. When speaking to responders I often pay attention to how much others are paying attention to me and to my message. I guess it's all part and parcel to being a student of situational awareness. I typically don't have much trouble capturing and holding the attention of an audience and often use what I have learned in my study of neuroscience to elicit strong connections with my audiences. But the one thing that disappoints me is the number of first responders who attend a full-day program (be it mine or someone else's) and don't bring any means of taking notes (e.g., pen and paper, iPad, laptop, recorder, *nothing!*). It often makes me ponder: Are they attending to learn or attending to be entertained?

Brain science has demonstrated that the most effective way to retain information is repetition. I spoke to this in the beginning of the book, warning you that I would be using repetition throughout this text to drive home important lessons. You may recall my sharing with you that research has shown the average person can only remember about seven pieces of unrelated information. In the course of a full-day program, instructors share hundreds of pieces of information and undoubtedly dozens of critical lessons. Research has further shown that as many as 90% of program attendees may forget as much as 60% of what they learn within the first three hours following an instructional program.

The remedy to this dismal recall situation is repetition. The problem is if attendees do not take notes, how can they apply the valuable learning principle of repetition? The solution I devised was this: I would take the notes for them. I have created that notebook—and you are holding it.

Thank you for coming along with me on this journey. I hope I have helped you learn to see the bad things coming in time to change the outcome.

Fig. 21–1. I have been blessed to be able to share my message with more than 30,000 first responders from across North America, South America, Europe, Asia, and Australia.

APPENDIX

SITUATIONAL AWARENESS FAQ

From the opportunities I've had to talk with responders about situational awareness and decision making, I have accumulated a list of frequently asked questions. I am confident the answers are contained within this text, but I also know that sometimes the answer may not be so obvious. So I've taken the liberty of providing you a list of 21 questions and answers. Additionally, I welcome the opportunity to get connected with you by e-mail or social media (I'm on Facebook, LinkedIn, Twitter, and YouTube) so you can't use the excuse that you didn't know how to reach me. Just do a Google search.

Q: Our department doesn't train on Mayday procedures at all. What can I do to make our chief and training division understand how crucial this is? We've had several close calls in our department, and I am really concerned.

A: You're really addressing two critical issues here. First is the need to have a procedure in place for a Mayday (and the personnel training on how to deploy the procedure). Second is the issue of the number of close calls your department has experienced. Close calls are the precursor—the warning signs, if you will—to catastrophes. Every time you have a close call it should invoke a thorough review of what happened, how it happened, and an analysis of how to prevent it from happening again (which may involve changes in policy or procedure), and most certainly training of personnel on how to prevent the near miss in the future. There are many parameters that need to be evaluated when there is a near-miss (or casualty) event, not just whatever went wrong at the moment of the incident.

As far as how to get your department leaders to embrace safety, I think the best approach is through education and subtle persuasion. Give them this book to read, or send them to my website (www.RichGasaway.com). I've made several training DVDs that can be used for a department drill. Use them to open up discussion about how your department could improve its safety practices.

Q: In the risk management area, how do you balance risk assessment with getting the job done?

A: This is a question I hear often. We are in a profession that has an inherent risk associated with what we do. This does not mean, however, that we need to be casual about how we assess our risks. (There's an old saying about love and romance that may apply here . . . that only fools rush in.) The fire service does a good job, for the most part, of training firefighters how to arrive and attack a fire from the interior. However, we don't do nearly as good a job training firefighters on what a no-go situation looks like. Thus, firefighters arrive and perform based on their training, which is most often (if not always) *go* inside and put the fire out. Many firefighters I have spoken to acknowledge they've never had meaningful training (or even a meaningful discussion for that matter) on what no-go looks like and when they should be defensive versus offensive in their attack. With a single script programmed in a firefighter's mind (go), it becomes predictable that will be the script deployed under stress.

As firefighters, we're tasked with trying to save lives from the ravages of fire. Unfortunately, not all lives are savable. There are conditions in which civilian victims cannot survive. There are also conditions in which firefighters cannot survive. Knowing what these conditions look like, and knowing what the precursor conditions leading up to lethal conditions look like, helps firefighters make better go or no-go decisions.

Q: What can I do, as a new officer, to enhance my skills in decision making?

A: In a word . . . Plenty! There are many ways to enhance your knowledge on a topic. Perhaps one of the best ways is experience. That's not the only way, yet there are plenty of firefighters out there who favor experience, as reflected in additional points awarded on promotional exams for years of service. This can prove to be a foolish (if not dangerous) way to promote. Why? Years on the job don't automatically correlate with the advancement of knowledge.

Maybe you've heard the saying that a person can have 20 years of experience, or they can have 1 year of experience repeated 20 times. Perhaps no truer words have been spoken when talking about some people in our profession. Experience does not equal job knowledge. Some readers aren't going to like hearing this (send your hate mail to Rich@RichGasaway.com), but others know it to be true and have seen it first-hand.

In defense of experience, research has shown that for a person to develop expert-level knowledge and performance in their chosen profession, it requires ten years of experience if the person practices their skill/art an hour a day, five days a week. Unless you have a severe arson problem in your town, you're probably not going to get that kind of practice.

So what do you do? Research shows that the brain stores and recalls vividly imagined experiences the same as if the experience were real. In other words, when your brain is storing experiences, it can't distinguish real experiences from vividly imagined experiences. It stores and uses both the same way.

This can be a tremendous asset for a young fire officer or aspiring officer. When you participate in training simulations or read case studies, line-of-duty death reports, or fire engineering articles, vividly imagine yourself as being involved in a real event and your brain will store the event for future recall. Throughout this book I reference how the brain uses the recognition-primed decision-making process. The vividly imagined experiences gained under nonemergency conditions can become part of the knowledge repository your brain can search through when you are under stress. In summary, it's possible to get 10 years of experience in less time by preloading experiences.

Q: You referred to a full-day training session on this topic. Where is the course information available?

A: You can get information about all my course offerings on my website (www.RichGasaway.com). Access the information about the programs by clicking the Educational Programs link on the left navigation bar. If you're interested in talking with me via email or in person, click the Contact Us link to access my email and cell phone number. Don't hesitate to contact me. I love talking about this topic with anyone who will lend me their ear.

Q: Is there common ground for the process between typical response scenarios and special hazard responses in industry?

A: I assume in the context of this question, the *process* you are referring to is the recognition-primed decision-making (RPDM) process. In that context, the answer is the process has been proven, through research, to be universal in a wide variety of high-risk, high-consequence decision-making roles including airline pilots, jet fighter pilots, submarine commanders, tank commanders, air traffic controllers, police officers, emergency room teams, surgical teams and, of course, fireground commanders.

I'm not aware of a specific study of RPDM within an industrial fire protection setting, but I can infer that the same principles would apply. Granted, the typical response scenarios found in an industrial setting are going to be different, but nonetheless many of them are common within industry and are the basis for training programs to develop expertise.

Q: How do you develop situational awareness skills in new recruits when the academy has programmed them to react instead of respond?

A: I'm not completely sure I fully understand the difference between *react* and *respond*, although I suspect the difference was crystal clear in the mind of this question's author. A person who develops a deep knowledge of a subject is in a favorable position to be a resilient problem solver.

I believe every academy should teach recruits how decisions are made under stress and how to develop and maintain their situational awareness. To be effective, situational awareness needs to be ingrained in every aspect of training, both in words and in actions. For example (and I was trained this way, so I'm as much a victim of this as any of you), we conduct training sessions for emergency medical services (EMS) providers where they are taught, repeatedly, to acknowledge they're wearing their personal protective equipment and the scene is safe.

However, the more times they say it without actually doing it, the more likely the script of saying it and not doing it will become the behavior displayed under stress. It's not enough to say you'll do physical tasks; you must actually perform the task to program muscle and cognitive memory in the scripts the brain runs under stress.

For example, let's look at how EMS responders train. During patient care scenarios, the team members are taught to say, "the scene is safe," yet most are never taught how to make sure a scene is actually safe. They're never taught how to evaluate the specific clues and cues that ensure safety. This can be a set-up for catastrophe because they're not assessing the clues and cues that ensure the scene is actually safe. This behavior, affirmed through repetition in training scenarios, programs a potentially dangerous script into the subconscious brain—a script that when played during a real emergency may find the responder saying "the scene is safe" without actually doing *anything* tangible to ensure that it is.

Q: I believe size-up is critical. Do you think you should let everyone know when a 360-degree size-up is complete? Maybe having a fourth benchmark, notifying everyone coming to the scene that a 360 size-up has been done instead of units just assuming it has been done?

A: In one word . . . Absolutely! You are correct in your statement about the importance of the size-up. The clues and cues captured during the size-up is the foundation of Level 1 situational awareness. Not only communicating the size-up has been completed, but also what critical clues and cues were noted helps develop the Level 1 situational awareness of everyone responding to the call.

Even if the next-in company arrives and sees something completely different from the first-arriving company, that's very important information about what has happened during that time. Issues with communications are among the most common contributing factors in near misses and casualties. Ensuring everyone is on the same page with a common understanding of what's going on, what's been done, and what needs to be done, may be one of the most important indicators of a safe incident outcome.

Q: How do you suggest to egotistical chiefs that the firefighters and the chiefs need to be on the same page tactically and strategically?

A: Sometimes ego eats brain and no amount of common sense will reverse the infliction of that disease. In my classes I like to use two analogies that help bring this point to life. The first is that of an orchestra where it's the job of the conductor to lead the musicians in the creation of, well, music. However, if the conductor and the musicians are not on the same page, there's not going to be music. There's going to be confusion, anger, frustration, and lots of noise being made, but it's not going to sound anything like music. This means as a group, there has been failure to accomplish the objective.

The second analogy I like to use is a football team where perhaps only the coach and the quarterback know the plays, but the rest of the team is left out of the proverbial loop. The quarterback calls the play in the huddle, and no one understands what it all means. There's no time to ask questions, though. The clock is ticking. So the huddle breaks, the players assumes their positions, and they do whatever their best judgment tells them is right. Oh my! You don't have to be much of a football fan to see in short order that's going to end in disaster. In orchestra or football, the results are rotten music or an embarrassing loss of the game. In our profession, the results may very well be a near miss or a firefighter casualty.

Q: I was recently promoted from lieutenant to fire chief. I have always made my decisions based on what was happening on the interior (where I was operating). How can I apply this tactic as the fire chief?

A: The short answer is . . . You can't. The clues and cues you processed in your decision making while operating on the interior are completely different from the clues and cues you process from the exterior. It's the equivalent of watching a football game from your seat inside the stadium versus standing beside your car in the parking lot and only hearing the cheers or boos of the fans inside. If you hear a cheer, you can assume something good has happened for the benefit of the home team, but you don't really know if that means a touchdown, a first down, or an interception. The same is true if you hear booing.

As a new commander (positioned on the outside, hopefully), your perspective is entirely different and the clues and cues you process are different as well. You have to look at things differently. For example, the color, volume, velocity, and density of the smoke, as observed from the exterior, is going to be different from the observations made from the interior. If you've not spent any appreciable time on the exterior prior to being promoted to a command officer role, you have some work to do to prepare yourself to be the big-picture commander.

One final word of caution to all newly promoted commanders—be careful! Under stress you can be drawn toward doing the tasks that you're comfortable doing. In this case, these would be hands-on firefighting skills. Avoid this temptation at all costs. You cannot be a hands-on firefighter and the big-picture commander at the same time. If you go hands-on, no one will be commanding the incident, and that can put you on a shortcut route to disaster.

Q: During your 20 years as chief, where did you command—in the seat, or in the street?

A: I have commanded from both positions. In my early years as a commander no one could have convinced me there was any other place to command from than in the street. I had to be in and among the action, which included plenty of hands-on work. I now know that was probably due to my hereditary stress reaction that drew me toward doing the tasks I was most comfortable doing. And for me, at that time, my comfort was definitely being in the action. I wasn't comfortable at all being on the sidelines as an apparent passive observer.

However, as time and the number of incidents I commanded passed, I was able to see how being close to the action almost automatically meant more distractions, interruptions, and impact on my ability to do what I was supposed to be doing (commanding the entire incident from a big-picture perspective). I started to realize I needed to be physically removed from the action. I can honestly say the transition from being a street commander to being a seat commander was challenging for me. Early on I found myself getting out of my vehicle to address problems and issues with personnel. Sometimes I made it back to the vehicle; sometimes I didn't.

Eventually, I was able to keep myself in the seat and developed the habit I believe reduced my distractions and interruptions and improved my decision making and situational awareness. If I was first to arrive, I conducted my 360 size-up before assuming my fixed and remote command position. If someone else was the first to arrive, I depended on their size-up. I found it beneficial to have an operations officer and/or safety officer who could be my forward set of eyes and ears to capture the clues and clues up close and personal, and communicate back to me the critical few pieces of information I needed to make good strategic decisions.

Q: What should response standard operating procedures (SOPs) look like?

A: Simply stated, response SOPs should mirror exactly what your department does at real emergencies. If I were to come in and assess your department, the performance I see on the street should look just like the SOPs, with some variation for unpredictable events that require resilient problem solving. If they don't match up, then there's a problem. Either the SOPs were written in isolation by well-intended bureaucrats and do not reflect the reality of the street, or the SOPs are solid best-practices that are being ignored by street practitioners because they've not been properly trained, don't understand the procedure, or are blatantly defying the standard.

When I look at a well-written SOP I should be able to envision it being performed by firefighters working under stress at an incident. It doesn't take a trained and experienced SOP detective much time to assess an SOP for its street reality potential. Response SOPs should be guidelines, which means they are not volumes in length. It's a common error of SOP developers to try to cover every possible scenario and give detailed instructions on what to do. That doesn't work so well. Keep your guidelines at the 30,000-foot view, and be sure to train personnel on how to be problem solvers when the situation varies from the SOP.

Q: What types of systems would you encourage firefighters to use to help support decision making in context?

A: The first—and perhaps the greatest—system for decision making in high-stress, high-consequence environments is your cognition (using your brain). As you heard me talk about how you make decisions under stress and what you need to have in place to make decisions, those are all skills and knowledge that can—and should—be developed in advance of needing them. Your brain is much better equipped to help you make good decisions when you experience similar events and store those activities for future recall. The solutions to problems the brain has memorized come much quicker than the solution to problems the brain has never seen before.

Here's a quick and easy exercise you can do to prove this. Ask one of the firefighters on your company to say the ABCs as fast as possible. On average that takes anywhere from 4 to 7 seconds (depending on whether it takes more than one breath). Now ask a different firefighter to say the ABCs backward as fast as possible. Unless they have memorized the alphabet backward (and I have actually met some people who have), the performance is going to be much slower. I would not be surprised if it took 4 to 7 minutes. From this elementary-school-style experiment you quickly see that the brain can recall information stored in long-term (subconscious) memory much more quickly. When novel (first-time) tasks are encountered, there is no long-term memory to draw upon. The brain must use short-term memory and conscious awareness to solve the problem—a process that requires much more brainpower, is much slower, and is far more likely to result in errors.

Second to using the brain, other helpful systems are those that help manage workload. One example is worksheets, which are great for helping a commander manage *retrospective* memory (the memory of everything that has been done on the emergency scene). Another example is checklists, which are helpful in managing a commander's *prospective* memory (the memory of everything that has yet to be done at the emergency scene). Of the two memories, research has shown that the prospective memory is the most vulnerable under stress.

Although there are more, the final system that I address in this response is the communications system, all-inclusive of hardware, software, nomenclature, cadence, scripts, commonly understood terms, and communications discipline. Communications ensures the success or failure of both relationships and emergency scenes.

Finally, it's been shown that verbal communications can consume a significant amount of your working memory to process and understand. This can be challenging when there are a lot of messages (e.g., heavy radio traffic). The more verbal clues and cues you must process, the more vulnerable you are to missing messages. If commanders are able to delegate the task of listening to the radio to someone else, that alone would substantially reduce the cognitive workload.

Q: Should fire academies incorporate situational awareness practicals into basic firefighting classes to get this mind-set established early in a firefighter's career?

A: Yes! In my fantasy world of firefighter training, the concepts of situational awareness and decision making would be the first lessons taught to new recruits (even before they are told where the bathroom is . . . that way they could use their situational awareness skills to find it).

All kidding aside, I believe with all my heart that building the deep knowledge of decision making and situational awareness is part of the solution to improving firefighter safety. Incorporating situational awareness components into every drill topic helps form long-term memory stores within a firefighter's brain. We should be thinking about the role of situational awareness in every task we perform, whether it's fighting a structural fire, working a code, or sweeping the apparatus floor.

Q: Is it detrimental to allow an EMS battalion chief (who has never fought a fire or ridden a ladder truck) to be a candidate for transfer to a fire battalion commander (BC) position?

A: I feel like I am leaning into a proverbial left hook with this response, and a copy of this answer is going to be printed and used for someone's benefit or detriment. Nonetheless, it's a good question and one I feel is worthy of a response. The skill sets for developing and maintaining the situational awareness of an EMS provider and a firefighter are very different, yet surprisingly similar. In their size-up of a situation, both have to capture clues and cues (Level 1 situational awareness), process those clues and cues into meaning (Level 2 situational awareness), and make reasonable predictions of future events (Level 3 situational awareness). But that is probably where the similarities end and the vast differences begin.

The stored experiences that drive the mental modeling under these two conditions are different. I use Michael Jordan as an example. Michael was, by all accounts, an amazing basketball player. An expert by every definition, he was nearly unstoppable in the application of his talent. But, Michael wasn't a very good baseball player. How could such a gifted athlete be so good in basketball and only mediocre in baseball? The complete brain science explanation is too long to get into here, but I think you get the point that a talented EMS commander would not automatically transition to become a talented fire commander.

In defense of the candidate, it's not impossible to do. There are plenty of examples of people who have taken talents earned in one discipline and applied them in another with great success. The resilient learning ability of the individual probably has as much to do with their success in a new endeavor as their previous experience. I will go as far as to say I am hopeful the EMS commander turned fireground commander would realize there is a steep learning curve and start preloading experiences to get up to that level. A mentor could watch out for the safety of the firefighters while the new commander is learning how to command (this is good advice for any new commander).

Q: Do you believe there is a disconnect when chief officers don't participate in line-level drills and training?

A: Yes, to a certain extent. To answer this, I go back to the response to the question where I used the analogies of the orchestra and the football team. Musicians rehearse some without the conductor present, and football players practice some of their plays independent of the coaches being present. However, at some point, the maestro and the coach must be present to observe the performance of the players and to give direction for improvements.

The chief officer who doesn't directly observe the performance of the crews is prone to having a flawed Level 3 situational awareness (predicting the future). Think of it as your firefighters playing the roles of actors in a movie. Each actor (or company of actors) has a role to play in making the movie turn out the way the director (incident commander) wants it to. For the director (commander) to accurately predict the future events at an incident, certain information about the actors (crews) need to be known. Some are obvious—crew size, location, and task assigned. Among the less obvious is an assessment of the quality of the crew and each member's ability to get the job done in the time frame the commander expects.

Let's face it (at the risk of sounding mean spirited), not all crews are created equal in their abilities. The *A-Team* crew can get much more done, more efficiently, and more effectively than the crew that looks and acts like they just arrived from the *Island of Misfit Toys*. The commander must know something about the abilities (and inabilities) of each team member. To do that, the commander must observe the actors in rehearsal of their parts. Being unfamiliar with the crews, the tasks being performed, and how long it takes to accomplish the tasks can seriously compromise the commander's Level 3 situational awareness.

Finally, I don't think command officers need to participate and perform the hands-on tasks that are company-level work. But observing the performance of the team members is beneficial in helping a commander understand the abilities of the firefighters, a critical component to command situational awareness.

Q: What do you say to firefighters who make the comment, "We go to fires. We know what we are doing," yet firefighters are continually getting hurt and killed on the fireground?

A: To address this bluntly, I borrow a quote from the comedian Ron White who notes, "You can't fix stupid." Or, maybe put more aptly, "You can't fix ignorance." What you describe here is one of six dangerous mind-sets that a firefighter or fireground commander can have that may affect decision making. These dangerous mind-sets include:

- Anti-authority. (The rules are not for me.)
- Impulsivity. (We must act now. There's no time to waste.)
- Invulnerability. (Nothing will go wrong. Those bad things only happen to other people, not me.)
- Resignation. (There's no use in me speaking up. No one listens to what I have to say.)
- Denial. (It's not really as bad as it looks. I've seen worse and handled it okay.)
- Macho (No one's going to tell me what I can do.).

This may be one of my biggest frustrations with reading the near-miss and line-of-duty death reports—many of the casualties keep happening in nearly the same predictable ways. It meets the definition of insanity: When a person continues to do the same thing over and over again, expecting the results to be different. In some respects, we are insane.

Q: I use time line prompting from dispatch to measure our progress. Is this good?

A: No, this is not good—this is exceptionally good, as long as your dispatcher is announcing the time benchmarks over a radio channel (or multicasting of multiple channels) so everyone on the incident scene hears the time benchmark. It's vitally important that every person operating at the emergency scene be prompted with time benchmarks to help maintain strong situational awareness.

One critical component of situational awareness is the changes that occur at the emergency during the passage of time. Under stress a person can suffer from *temporal distortion*. That's just a fancy way of saying you lose track of time. When your senses are being bombarded with stimuli and you are under stress, your timing mechanism can get out of whack.

Ensuring everyone has the same awareness of the passage of time helps keep everyone attuned to what is happening at the emergency scene in the context of time. This is vitally important. For everyone who has their dispatch marking the passage of time with benchmarks, my hat's off to you.

Q: At our last fire, a captain who was the officer first on-scene assumed command. When the department chief arrived, only one lightly staffed truck was on scene and the chief grabbed a hand line and began protecting exposures. After additional resources arrived, the chief passed off the nozzle and assumed command. Do you think this is appropriate behavior for someone of that rank?

A: I am going to reframe one component of your question, because rank is not the issue I want to address. What I want to address is the role the person was filling when the nozzle was deployed. I am often told of incidents where the person who was allegedly serving as the incident commander was performing some hands-on task at the incident scene, and as a result missed some really important information that impacted crew safety.

When the person who is telling me about these incidents is the same person who was supposed to be the big-picture commander, the story is often accompanied by a slew of excuses offered to justify why the person had to perform hands-on tasks. Here's what I can tell you with confidence: The act of commanding an incident requires a great deal of cognitive energy and very strong situational awareness. It requires the complete and entire attention of the person in charge to be in that role.

Anytime the person who is in charge goes hands-on because of ______________ (fill in any excuse you want to put on the line), that person is no longer the big-picture incident commander, period. You can't do both things (well) simultaneously. In fact, as much as this is going to come to stir controversy to say it (send your rebuttals to Rich@RichGasaway.com), there's a slew of research that demonstrates the human brain cannot multitask conscious activities performed in a stimulus-rich environment.

The instant a commander goes hands-on, he or she exponentially increase the likelihood of error. Research has shown that a person whose work is interrupted will make, on average, 50% more mistakes and is likely to miss up to 50% more visual and audible clues and cues. A commander who goes hands-on is dancing with the devil.

Perhaps among the most insidious outcomes of a commander going hands-on comes when he or she gets away with it (the incident does not result in a near miss or a casualty). This lack of *consequence* can falsely build a commander's *confidence*. The more times this happens, the greater the confidence they gain, and the more they believe hands-on work while commanding to be acceptable. Don't confuse competence with being lucky.

Q: How do you feel about the incident commander staying in the buggy versus being in front of the fire building?

A: I addressed this some with an earlier response, so I will try not to repeat myself here. Some of the greatest challenges a commander faces in developing and maintaining situational awareness relates to interruptions, distractions, communications, attention management, and workload management. The results of my research make a strong argument that the barriers to situational awareness are improved when the commander is situated in a remote location where a visual fix can be maintained on the broad incident scene.

Does this have to be in a vehicle? I guess not. But being in a vehicle surely creates an environment more conducive to controlling distractions and interruptions; monitoring radio communications (i.e., never missing important radio messages from at-risk companies); improving the command environment (even with simple things like controlling the lighting and temperature and getting out of the rain or snow); and improving workload management by facilitating the use of worksheets, checklists, preplans, and computer-aided building information or personnel tracking software.

Q: You mentioned victim survivability and firefighter survivability. In the same conditions in a fire, when is it time to write off survivability?

A: I speak to the survivability of civilians and firefighters in the context of windows of opportunity for survivability. When a fire is small, the window of opportunity for a civilian victim self-rescue is wide open. Short of the civilian occupant being physically or mentally incapable of self-rescue, a person inside a building with a small fire will generally be able to get out.

We see that all the time as the vast majority of the structure fires we fight do not have victims inside. As the fire continues to grow and the heat and smoke build up and bank down toward the floor, the window of survivability for civilian victims decreases to the point where the only way they survive is for a firefighter to enter the environment with protective equipment and rescue the victim. Because skin begins to melt at 160 degrees, the window of time where a victim can survive at the hands of a firefighter rescue is very small.

When a victim is exposed to 200+ degree temperatures, skin begins to sear and the injuries can quickly become incompatible with life. Burn center staff readily admit that a fire victim with third-degree burns over 70% of the body has a very low probability of survival. The elderly and children are the most vulnerable to succumb to severe burn injuries. Firefighters, thanks to their protective equipment, are able to easily survive and work in environments that would kill an unprotected civilian victim, so a firefighter's window of survivability is different. However, don't be fooled into a false sense of security. Different does not necessarily buy you much more time. The increase in hydrocarbon-based furniture and contents in structures exponentially accelerates heat, smoke, and fire growth curves, resulting in rapidly closing windows of survivability (for both civilian victims and firefighters).

That's not to say there aren't clues and cues that the windows are closing or have closed, because there were plenty of indicators in most cases I've reviewed. Imagine if there was a crystal ball that you could look into and see what the fire conditions were going to be in 5 minutes. That would be an amazing discovery and would save countless firefighters' lives. If only such a thing existed. Well, it does. It's the smoke. The smoke is the crystal ball and the indicator of the future. The color, volume, velocity, and density of the smoke can tell you so much about what's burning, where it's burning, how deep the fire is, how hot the fire is, and whether structural components may be burning (and preparing to collapse). But to use the crystal ball, one must have situational awareness.

Q: If you equate tacit knowledge with experience, how would one compensate for a lack of experience in becoming a good fireground commander, even if equipped with a solid decision-making capability?

A: Don't be lulled into believing there is only one way to get experience (in the street). You can pre-load your experiences with realistic incident simulations, case studies, near-miss reports, line-of-duty death reports, and magazine articles. For example, you may have never responded to a brush fire sparked off by a downed electrical line. However, if you have read the line-of-duty death report where a firefighter was mortally wounded when he stepped on a downed high-voltage line while extinguishing a routine brush fire, you can pre-load that experience and use that knowledge to help improve your situational awareness when operating at all future brush fire calls, even though you've never actually experienced a situation similar to this in real life.

INDEX

A

E

F

G

K

L

M

N

O

P

R

S

ABOUT THE AUTHOR

Richard B. Gasaway, PhD, has served 33 years as a public safety professional for six organizations in West Virginia, Ohio, and Minnesota. He has held positions of firefighter, EMT, paramedic, lieutenant, captain, training officer, assistant chief, and fire chief.

In addition to his lifelong commitment of service to others, Dr. Gasaway has a second passion—understanding how the brain works and relaying those lessons in meaningful ways for the benefit of first responders.

As of this writing, Dr. Gasaway's programs on situational awareness and decision making have been presented more than 2,500 times to some 30,000 first responders from North America, Europe, Asia, and Australia. He has contributed to more than 350 journal and website articles, books and book chapters on the topics of first responder leadership and safety.

Dr. Gasaway's contributions and dedication to sharing his knowledge with public safety providers has led to him being widely considered as one of the foremost authorities on first responder decision making and situational awareness under stress.

Contact information:

Gasaway Consulting Group

www.RichGasaway.com

Situational Awareness Matters

www.SAMatters.com

Email: Rich@SAMatters.com

Phone: 612-548-4424